AF786643

FRICTION AND ANTIFRICTION MATERIALS

PERSPECTIVES IN POWDER METALLURGY
Fundamentals, Methods, and Applications

Volume 1 (1967)
New Methods for the Consolidation of Metal Powders

Volume 2 (1967)
Vibratory Compacting: Principles and Methods

Volume 3 (1968)
Iron Powder Metallurgy

Volume 4 (1970)
Friction and Antifriction Materials

Volume 5 (1970)
Advanced Experimental Techniques in Powder Metallurgy

PERSPECTIVES IN POWDER METALLURGY
Fundamentals, Methods, and Applications

Editors:

Henry H. Hausner
Adjunct Professor
Polytechnic Institute of Brooklyn
Consulting Engineer

Kempton H. Roll
Executive Director
Metal Powder
Industries Federation

Peter K. Johnson
Assistant Director
Metal Powder
Industries Federation

Volume 4

FRICTION AND ANTIFRICTION MATERIALS

with a Foreword by Erik Navara and Václav Procházka
and an Introduction by Henry H. Hausner

PLENUM PRESS · NEW YORK–LONDON · 1970

ISBN 978-1-4684-3017-2 ISBN 978-1-4684-3015-8 (eBook)
DOI 10.1007/978-1-4684-3015-8

Library of Congress Catalog Card Number 74-127937
SBN 306-37684-9

Softcover reprint of the hardcover 1st edition 1970
A Division of Plenum Publishing Corporation
227 West 17 Street, New York, N.Y. 10011

United Kingdom edition published by Plenum Press, London
A Division of Plenum Publishing Company, Ltd.
Donington House, 30 Norfolk Street, London, W.C.2, England

ACKNOWLEDGMENTS

The editors gratefully acknowledge permission to reprint the following articles:

Sintered Metal Brake Linings for Automotive Applications, by E. W. Reinsch, *Progress in Powder Metallurgy,* 1962, pp. 131–138.

A Contribution to the Investigation of Dry Friction of Sintered Steel, by V. Procházka and V. Miskovic, *Kovove Materialy,* 1964, pp. 408–418.

The Influence of Silica and Alumina Additions to Sintered Iron-Based Friction Materials, by V. Procházka and E. Navara, *International Journal of Powder Metallurgy,* April 1966, pp. 39–47.

Porous Metal Bearings, by V. T. Morgan, in: *The Principles of Lubrication* (A. Cameron, ed.), John Wiley & Sons, New York, 1966, pp. 542–559.

Lubrication of a Porous Bearing—Stokes' Solution, by D. D. Joseph and L. N. Tao, *Journal of Applied Mechanics,* 1966, Paper No. 66-APM-KK.

Lubrication of a Porous Bearing—Reynolds' Solution, by C. C. Shir and D. D. Joseph, *Journal of Applied Mechanics,* 1966, Paper No. 66-APM-IJ.

Selecting the Right Lubricant for Self-Lubricating Bearings and Parts, by J. J. Scott, *Progress in Powder Metallurgy,* 1963, pp. 127–134.

Sintered Bearings, by A. Adler, *Precision Metal,* May 1969, pp. 51–53.

Effect of Copper Additions on the Bearing Properties of Sintered Iron–Graphite, by A. Domsa, *Neue Hütte,* 1964, pp. 186–188.

Production and Properties of a New Porous Bearing, by H. Youssef and M. Eudier, in: *Modern Developments in Powder Metallurgy,* Vol. 3 (H. H. Hausner, ed.), Plenum Press, New York, 1966, pp. 129–137.

Cost Structure for Self-Lubricating Bearings, by O. Wiberg, *Powder Metallurgy,* 1965, pp. 231–233.

The following articles are adapted from *Soviet Powder Metallurgy and Metal Ceramics,* the English translation of *Poroshkovaya Metallurgiya,* published by Consultants Bureau, New York.

Effect of Graphite Content on the Antifriction Properties of Metallographic Materials, by L. I. Pugina, I. M. Fedorchenko, and N. E. Ponomarenko, 1965, pp. 53–58.

Experience in Manufacturing Iron-Base Sintered Friction Alloys, by L. P. Lebedeva, B. T. Krysin, Ya. V. Kolpakov, L. N. Ignatov, V. A. Mikhailovskii, G. G. Smirnov, and M. V. Tsytsenko, 1965, pp. 96–102.

Studies of Friction Materials, by I. M. Fedorchenko, I. I. Panaioti, G. M. Derkacheva, I. Ya. Dzykovich, and G. N. Gordan, 1965, pp. 54–57, 65–68.

Effect of Sizing Allowance on the Surface Quality of Iron–Graphite Bushings, by D. Z. Yurchenko, 1966, pp. 62–66.

The following articles are adapted from the *Proceedings of the Second International Powder Metallurgy Conference,* Tatra, 1966:

The Influence of Structure and Metal Additions on the Properties of Iron–Graphite Sintered Materials, by E. Strobel and H. Rebsch, Vol. III, pp. 161–178.

Iron-Based Sintered Antifriction Materials for Heavy-Duty Service, by I. M. Fedorchenko, L. I. Pugina, V. V. Pushkarev, J. F. Shevchuk, and V. S. Ageyeva, Vol. III, pp. 153–159.

The Influence of Copper on the Properties of Sintered Iron–Graphite Friction Materials, by V. Procházka, E. Navara, V. Miskovic, and K. Kosta, Vol. III, pp. 179–196.

The Bearing Performance of Sintered Metal Bearings, by G. C. Pratt, Vol. III, pp. 29–52.

A New Dry-Running Bearing Material, by Z. Ministr and J. Priester, Vol. III, pp. 71–86.

The following articles are taken from the *Proceedings of the Seventh Soviet Scientific Conference on Powder Metallurgy,* Yerevan, 1964:

Some Recent Advances in the Manufacture of the Friction Sintered Material Grade FMK-11, by B. T. Krysin, L. N. Ignatov, and Ya. V. Kolpakov, pp. 257–265.

New Sintered Friction Materials, by I. M. Fedorchenko and V. M. Kryachek, pp. 186–189.

Introduction

Powder metallurgy literature in the English language includes a large number of books and several thousand articles in various journals. The rate of growth of this literature increases from year to year. It covers well the whole field of powder metallurgy – materials, processes and products – with two exceptions: friction and antifriction branches of powder metallurgy.

This lack of information has nothing to do with scientific or technical considerations, and definitely has nothing to do with lack of initiative in the development of these materials. The industry concerned with the production of friction and antifriction materials is continually developing new products and techniques and production is steadily growing. However, most companies working on these materials regard their experiences and new advancements as "proprietary" and, for competitive reasons, are not interested in publishing in the technical literature except for very perfunctory and usually highly commercial papers. Very little work on friction and antifriction materials is going on in independent laboratories and university laboratories, although fundamental studies in this field offer very interesting aspects.

What has been said above about proprietary ideas refers principally to Western countries, where competition between privately owned industries is encouraged. It does not apply to that part of the world where industry is owned by the state and where industrial competition therefore does not exist. In the latter case there exists a greater incentive to publish and to disclose what might otherwise be considered proprietary information. This is certainly the case with friction and antifriction materials.

After this explanation, the reader will not be surprised to find that more than 50% of the content of this book is translated from the literature in USSR and CSSR where the problems of friction and antifriction materials are more openly discussed and described, and where their solutions are not regarded as proprietary by the industry.

Being fairly well acquainted with the powder metallurgy literature in various parts of the world, I suggested that articles on friction and antifriction materials from both East and West should be collected. This is how Volume 4 of Perspectives in Powder Metallurgy originated. The 24 chapters of this book may contain some duplications which were difficult to avoid or eliminate because of translation problems. Nevertheless, we sincerely hope that the readers will enjoy this first English language book on friction and antifriction materials.

New York, N. Y.

Henry H. Hausner
Consulting Engineer

Foreword

Erik Navara and Václav Procházka

Laboratory of Metallurgical Technology
Technical University
Košice, Czechoslovakia

Friction and antifriction materials produced by powder metallurgy methods have long held a permanent place in materials engineering. They have experienced rapid growth in production and technological development, leading to improved properties and increased load-carrying capacities. But, despite this, the literature dealing with these materials is surprisingly inadequate. There are many patents available, covering almost every process and raw materials development. Many books on powder metallurgy have at least one chapter or two devoted to these materials. And papers on this subject are regularly presented at symposia on powder metallurgy. What becomes apparent is not a shortage of technical or scientific publications, but a lack of coordinated information that covers the subject thoroughly and points out the significant contemporary developments in fundamental research and commercial application.

Even books published thus far on P/M friction materials are uncommon and outdated. In 1960, W. Cegielski and W. Rutkowski published "Sintered Bearings," a well written book in the Polish language. Now, however, this is partially obsolete because many new research, development and production results have appeared since. A similar survey of the contemporary state of development, production, and application of friction materials was given by V. Dufek in his "Metal-Ceramic Friction Materials," a book published in the Czech language in 1965. But this was limited in scope and in distribution.

The purpose of this book, another in a continuing series of "New Perspectives in Powder Metallurgy" and containing 24 selected papers by authors from many different countries, is to fill the gap. Hopefully, it will inform the reader of the many achievements that have taken place in this important sector of powder metallurgy technology.

The first part is devoted to sintered friction materials. It is divided into several categories according to the base metal, application, and properties required. The base materials covered are iron and copper.

Copper, which almost invariably contains tin, was developed much earlier. Because copper does not react with graphite, which is a very effective wear-reducing additive, the maintaining of consistent properties in manufacturing practice is relatively simple. Sintering atmosphere control requirements are easy to fulfill. The basic disadvantages of copper-base materials are their high material cost and the low permissible operating temperature which is due to the low melting point of tin bronze.

Iron-base materials are generally used for heavy-duty conditions. However, their manufacture is more complex. Sintering temperatures are considerably higher, and a number of diffusional processes and chemical reactions take place which have a profound influence on the final properties. The carburizing of iron is of primary importance, and this, together with a number of other reactions, is discussed in several chapters. It becomes apparent that the production of sintered iron-base friction material requires close control of a large number of variables if quality is to be maintained.

It should be stressed that production of friction materials requires technical skill, experience, and modern equipment. Because properties depend greatly on production conditions, a change in manufacturing technology designed to improve a product for one application may impair its performance in another. Most production engineers have observed that an apparently minor change in raw materials or technology can lead to significant changes in the performance of the final product. Graphite is an especially intriguing factor, as B. T. Collins and C. P. Schneider point out in their excellent paper, "Sintered Metal Friction Materials," presented at the 1965 International P/M Conference in New York.* The addition of copper, for instance, affects the structure of iron-base matrix and, consequently, the friction and wear characteristics of the product. The selection of a suitable abrasive for dry application is no simple matter either, and its behavior during sintering has to be considered. Finally, several compounds which are sometimes used as stabilizers of the friction coefficient at elevated temperatures (e.g., barium sulfate) perform a complex role in the process which is not yet fully understood even though the merit of the additives has been proved empirically.

It is, therefore, not surprising that the conclusions of some chapters are at variance with one another, and sometimes even controversial. Depending on the type of performance, whether it is dry or under a lubricant, or operating at high or low temperature, different requirements are considered essential. These naturally have an effect on the preferred type of structure and, consequently, on the variables used in the manufacturing process.

Sintered antifriction materials and final products manufactured from them present substantially different problems for both investigators and production engineers, depending on the type of the product. Generally speaking, there are three main types of sintered antifriction products.

1. Porous self-lubricating bearings or bushings, where interconnected porosity control is of primary importance. The pores, which must form an interconnected network, are filled with a high-grade oxidation-resistant lubricant. In performance, this lubricant is squeezed out onto the sliding surface at the points of higher loading and sucked back into the pores at regions of lower loading. Thus a circulation is maintained, which is essential for the proper performance of self-lubricating P/M bearings.

2. Sintered linings of plain steel-backed bearings operating under oil-pressure lubrication. In this case, the porosity is minimal or eliminated. At first, powder metallurgy was specified to

---

* "Modern Developments in Powder Metallurgy," Vol. 3, Plenum Press, New York (1966), pp. 160-165.

overcome difficulties in casting technology. Later, however, several types of linings were developed which could not be manufactured by any technique other than P/M. Binary copper—lead sintered linings are a well known example of such a change in technology. These and porous linings infiltrated with a softer low-melting bearing alloy illustrate the versatility of P/M processes.

3. Dry bearings of linings containing solid lubricants within the pores of a sintered metallic matrix. One of the most successful products of this type is the Glacier DU dry bearing. It has a tin bronze sintered layer, whose pores contain a mixture of PTFE and lead powders pressed by rolling. This and similar types of dry bearings are capable of long performance under relatively light loads without any maintenance.

Most types of antifriction materials are covered in the second part of the book. Interesting problems and the latest research and development findings are also presented.

Contents

FRICTION MATERIALS

ANTIFRICTION MATERIALS

FRICTION MATERIALS

Chapter 1

The U.S. Friction Materials Industry

Ben T. Collins
Director of Research and Product Development
Wabash Division, Raybestos-Manhattan, Inc.
Crawfordsville, Indiana

The process for making a bonded bimetallic sintered metal articles is very old. In 1885 Canda [1] received a patent disclosing a process for sintering a metal bearing lining to a tinned bearing shell. In the early 1920's several people became active in producing sintered metal bearings and similar mechanical articles. By the late 1920's the use of sintered metal bearings had been adopted in the automotive and allied industries. From there, they progressed into the refrigeration and electric motor industries.

The sintered metal bearing industry was the progenitor of the sintered metal friction material industry. In 1929 Guay [2] disclosed that a composition patented by Gibson [3] in 1916 was a friction material—copper base with minor amounts of lead, tin, and graphite with or without magnesia, talc, or chalk.

Two of the pioneers to commercialize the new industry in the early 1930's were J. S. Fisher and S. K. Wellman. Late in the 1930's energy input and temperatures were getting so high that organic-base friction materials were starting to fade in service and give short life which resulted in accelerated use of the sintered metal friction material. By this time sintered metal friction materials were being used in earthmovers, trucks, tanks, and airplanes. The advent of World War II with its resultant demands for large quantities of heavy-duty friction material in military vehicles and aircraft gave a big impetus to the growth of the new industry. Natural growth of the national economy plus an economic push to put more and more energy through the same size clutches and brakes has continued the growth of the industry.

An idea as to the size and growth rate of the industry can be obtained from the following sales figures: $20 million in 1960, $24 million in 1962, and about $33 million in 1968.

Sintered metal constitutes a minor portion of the overall friction material industry, only about 15%. Competitive materials are cork, steel, resilient fibrous compounds, compounds with resin and rubber binders, and woven asbestos materials. Metallic fillers may be used in the latter two types of materials. These materials will be replacing sintered metal in some applications and will be replaced by sintered metals in other applications.

There are two major types of applications for friction material: dry and in oil. In general the oil applications are clutches while the dry applications are either clutches or brakes. Until about 1950 all of the applications were dry, but at that time sintered metal friction material was adopted for automatic transmissions in automobiles. In spite of the lower coefficient of friction and the greater complexity of design for the oil applications the increased life makes it economically desirable for the equipment manufacturers to convert from the dry applications except where original cost is the primary consideration. About 75% of the original equipment parts made today are used in oil.

The majority of the sintered metal friction material applications can be found in the following tabulation:

A. Dry clutch
   Mild duty: Farm tractors, machine tools, aircraft controls, and highway trucks
   Moderate duty: Blank and punch presses, and industrial material movers
B. Dry brake
   Moderate duty: Automotive and blank and punch presses
   Severe duty: Aircraft

C. Wet clutch
Mild duty: Light automatic transmissions, machine tools, and tractor steering clutches
Severe duty: Power shift transmissions in tractors and military vehicles

D. Wet brake
Moderate to severe duty: Off-highway trucks and military vehicles

Normally the differentiation between a clutch and a brake can be made on the basis of the duration of energy conversion; the clutch will work for about one second while a brake will work from one to about thirty seconds.

There is a big difference in the finishing operations for parts performing dry and in oil. For parts operating in oil it is necessary to groove the surface. The purpose of the grooves is to expedite the removal of oil from the interface and raise the coefficient of friction. Parts operating dry will only have a few coarse, if any, grooves. The grooving patterns will range from a spiral to radial grooves with practically all possible intermediate variations or combinations.

Compositions of sintered metal friction materials have changed, reflecting changes in applications, increased energy requirements, and the expansion of knowledge. Originally all friction compounds were copper-based, but now in addition to improved copper-base materials being used in all applications iron-base compounds have been developed for moderate to severe duty dry applications. In developing friction compounds it should be remembered that they are only half of the friction couple and that compatibility of the opposing member is equally important. True evaluations can only be obtained when the two members are conditioned to each other.

Three typical early compositions for dry clutches and brakes are:

73% Cu, 14% Pb, 7% Sn, and 6% graphite.
70.9% Cu, 10.9% Pb, 6.3% Sn, 4.5% sand, and 7.4% graphite
62% Cu, 8% Fe, 12% Pb, 7% Sn, 4% sand, and 7% graphite

Some typical current compositions for specific applications are:

Dry clutches and brakes:	75% Cu, 6% Pb, 7% Sn, 5% graphite, 4% MoS_2, and 3% feldspar
	56% Fe, 14% Zn, 18% graphite, 8% chromite, and 4% SiC
Aircraft brakes:	63% Fe, 4% chilled iron grit, 3% iron silicide, 2.5% PbO, 4.5% Sb, 6% SiC
Oil clutches and brakes:	74% Cu, 3.5% Sn, 2% Sb, 16% graphite, and 4.5% galena

The large decrease in lead content in current compounds should be noted; but some recent development work with iron-base compounds with high lead content has shown promising results for dry applications. Another pronounced change in compounding is the wide choice of abrasive materials that have supplemented the use of plain silica sand.

In spite of the many technological advancements in the powder metallurgy industry, friction material compounding still remains an art. A prior publication [4] gives a general review of ingredients and enumerates some of the problems that can be encountered.

There have been practically no technological advancements in the processing equipment since the conception of the industry, only minor refinements and increases in size. Some unique processes and products have been commercialized.

The most important development is primarily concerned with an improvement in the strength of the steel core. The ultimate in steel core strength has been an SAE 4140 steel hardened to Rockwell C 30 to 40, plus a nitrided case at least 0.020 in. deep. Better friction material performance has been obtained by prior hardening of the steel followed by a low-temperature sintering cycle rather than heat treating the assembly after sintering.

Another unique development has been a technique for sintering friction material to a conical surface on the inside surface of a ring.

Within the past decade the energy capacity of oil clutch friction materials has been doubled. There was a minimal increase in coefficient of friction while the energy capacity was being improved. In about the last five years the energy capacity of aircraft brake materials has been increased about 25%; this improvement has occurred without the loss of any of the other desirable characteristics.

It is impossible to predict the future of the industry, but it can be stated that it will expand in a direct ratio to the increase in energy capacity of the sintered metal friction compounds. The one material which would produce great increases in use would be a compound for very-high-speed oil clutches. These clutches would be coupled to turbine engines with a tremendous saving in size. A field which has been explored very little is the operation of friction material on friction material. Such a friction couple could lead to some very startling developments. Although sintered metal automotive brake lining for the drum-and-shoe brake has proved to be very successful for NASCAR racers, police cars, and other severe duty applications, it doesn't appear that it will ever become standard equipment on automobiles.

References

1. Canda, U. S. Patent 313,916 (March 17, 1885).
2. Guay, U. S. Patent 1,700,493 (January 29, 1929).
3. Gibson, U. S. Patent 1,177,407 (March 28, 1916).
4. Collins, B. T., and Schneider, C. P., "Modern Developments in Powder Metallurgy," Vol. 3, Plenum Press, New York (1966), pp. 160-165.

Chapter 2

Sintered Metal Brake Linings For Automotive Applications

E. W. Reinsch

Delco-Moraine Division
General Motors Corp.
Dayton, Ohio

The doctor, the housewife, the barber, the factory worker, most of you who are reading this paper—the average driver—takes his automobile for granted. It starts, it stops, it requires gas—it furnishes transportation. You are aware that power is needed to make the car go, but you probably do not realize that power is also required to make the car stop. This power requirement can be surprisingly high. For example, to stop a 5000-lb automobile from 100 mph in 10 seconds requires the brakes to absorb about 300 horsepower. This is more horsepower than is available in the majority of engines in service today.

It is the function of the brakes to absorb the energy of motion, convert it into heat, and dissipate the heat to the atmosphere. If heat is put into the brakes faster than it can be removed, the linings eventually reach a temperature where the coefficient of friction falls rapidly and causes the brakes to lose their stopping ability. This loss in effectiveness is termed "brake fade." As the brakes cool below the fade point, they regain their normal output. Repeated subjection or prolonged exposure to excessive temperatures can cause permanent damage to linings. This damage may manifest itself in brake performance as overeffectiveness, erratic action, rapid lining wear, and noise.

Brake fade may result from a variety of types of driving: an emergency vehicle making a high-speed run which requires hard and frequent brake applications, mountain driving when brakes are used to hold down car speed, or many stops in very rapid succession from slow speeds. In many instances linings may go through several partial fade periods without the driver being aware that his linings have been overheated.

The elimination of brake fade can be accomplished by either of two methods: (1) Make provision for more adequate cooling, or (2) use a lining material that will not be sensitive to heat.

Delco-Moraine has chosen the second approach to a solution; namely, develop a lining material which will retain effectiveness regardless of temperature! The use of metal powders and powder metallurgy techniques appeared to offer an approach to the fabrication of such a material. Our organization is uniquely adapted to engage in a development program on metallic linings since we have had over 25 years of experience in the field of powder metallurgy and in the design, manufacture, and testing of brakes.

The Duo-Servo Brake

The type of brake used in the work discussed in this paper is the duo-servo brake which is standard on the majority of American cars. A schematic drawing is shown in Fig. 1. This is a drum-type brake utilizing a single wheel cylinder and a primary and secondary shoe. It is self-energizing in both directions.

For a forward stop, drum rotating counterclockwise, hydraulic pressure from the master cylinder to the wheel cylinder forces the top of the primary shoe against the drum. Friction of the lining tends to rotate the shoe and the resulting force is applied through the adjusting strut to the secondary shoe. The secondary also tends to rotate but is restrained by the anchor pin. Thus there is a continuous build-up of forces from the leading edge of the primary to the trailing edge of the secondary. For a conventionally lined 11-inch-diameter brake, 1 lb of force at the wheel cylinder will produce 7.3 in.-lb of torque from the primary and 28.4 in.-lb from the secondary for a total of 35.7 in.-lb.

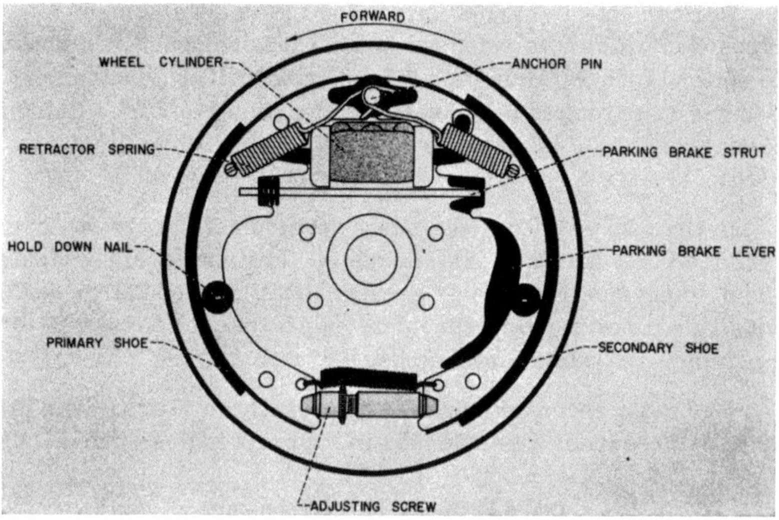

Fig. 1. Duo-servo rear brake.

History

Delco-Moraine's original interest in metal lining goes back to January, 1937, when a car test was made of a cast copper-base metal lining. While this material was unsatisfactory, the test marked the beginning of a development program that resulted in the acceptance of a metallic brake lining some twenty-one years later.

A major advance in the search for an antifade material took place several years ago when a high-graphite iron-base powder mix was compounded. Car tests showed this mix, out of over 400 previously evaluated, to have excellent fade resistance. However, other properties, such as noise and wear rate, were such that the material was not of optimum character for vehicle use.

In an effort to combine the good fade resistance of metallics with the advantages of conventional lining, a number of tests were run in which metallic and conventional materials were combined. Among these combinations were metallic front–conventional rear brakes, conventional primary–metallic secondary, and arrangements where metal and conventional lining segments were on the same shoe. None of these gave a completely satisfactorily performing brake and it was decided to concentrate on full metallics, primary and secondary, front and rear.

Refinements in earlier mixes led to the development of a composition which was released in 1958 by Chevrolet as a heavy-duty option, such as for police and taxicab service. This material was made in segments and riveted to the shoe rim. The lining did a commendable job in reducing fade under conditions of severe braking.

In the course of the continuing development program, it was determined that better overall brake performance could be obtained if the primary coefficient was greater than the secondary coefficient. This meant that different materials had to be considered for the primary and secondary linings.

Our most recent material, significantly improved over that originally produced, was released in 1960. It is designated DM81.

DM81 Metallic Lining

Figure 2 shows a brake assembly with DM81 sintered metal lining segments welded to the shoe rim. This lining is now used as an option on all Chevrolets, including Corvette, Corvair, and Chevy II.

The segments are duplex in structure—a friction surface and a dense backing material. The backing is used to increase strength of the segment and to facilitate welding.

Segments are made using conventional powder metallurgy techniques—mixing, briquetting, and sintering. Briquetting is slightly different than usual in that a double fill is required to produce the two-layer part. The duplex structure presented some problems in that it is necessary that the two layers change dimension to the same degree during sintering so that there is no separation and correct curvature is maintained.

In order to insure proper performance of the final product, a very comprehensive quality control system has been initiated. This includes pre-production evaluation of the individual ingredients and of the mix, check of segments during processing, and final inspection of the finished parts. Among the checks made are:

1. Dimension
2. Physical strength
3. Chemical analysis
4. Metallographic analysis

Fig. 2. Brake assembly with DM81 sintered metal lining segments welded to the shoe rim.

5. Frictional properties
6. Wear rate
7. Weld strength

Furnace temperature, speed, and atmosphere are also carefully controlled.

Laboratory Testing

Development of a friction element follows a path with many, many turns and more than its share of crossroads. An unlimited number of ingredients in a wide range of particle distributions, shapes, and sizes are available. The proportions in which these ingredients can be mixed are countless. Added to this are the variables produced by changing the density and/or the sintering conditions.

We have attempted to use a scientific approach in our development program, but the promising mixes have been few and often far between. As of October 1, 1961, we have investigated over 1600 compositions. The number has increased considerably since that time.

Metal lining segments differ from most sintered metal parts in that their most important single property is friction. The coefficient of friction is determined by the chemistry and by the metallurgical structure, which, in turn, results from the ingredients and briquetting and sintering techniques.

The basic tool used for measuring the coefficient of friction is a laboratory brake lining test machine as shown in Fig. 3. This machine is an absorption dynamometer incorporating a drum rotating on a horizontal axis. The test friction specimen rides on the rubbing surface on the bottom vertical centerline of the drum. Load is applied to the specimen and the resulting frictional force, which is a horizontal force acting tangent to the drum surface, is measured. Knowing the frictional force F and the load on the specimen N (normal force), the coefficient of friction u can be calculated from the relationship $F - u N$. Drum temperature, drum speed, and load are controllable.

Four of these machines are utilized in our plant, three for development work and a fourth for quality control.

One of the problems inherent in any laboratory test program is the correlation of results with actual performance. Thus, a proper laboratory test procedure must be available in order to accurately evaluate compounds. While we feel we have such a procedure, there is ample room for improvement. The fact that the primary and secondary linings have different coefficients and that the duo-servo brake is self-energized further complicates the picture. At most, the laboratory machine does only a screening operation – final tests are made on the car. However, the better the screening, the fewer the number of car tests. We would like to be able not only to separate the pearls from the peanuts, but also to differentiate between the types of pearls.

Car Tests

The evaluation of brake linings on a vehicle covers the following major items:

1. Effectiveness. How much output do the brakes have for a given pedal force? Is the output always consistent? Does the output change during a given stop, or after x miles of driving?

Fig. 3. Brake lining test apparatus.

2. F a d e. How do the linings react to heat?
3. S t a b i l i t y. Do the brakes provide straight, smooth stops from all speeds at all deceleration rates under all driving conditions? How do they respond to dampness, wet roads, and high and low ambient temperatures?
4. W e a r. What is the lining life under various driving conditions?
5. N o i s e. Under what conditions of driving are the brakes noisy?

Lining evaluation tests are conducted on our fleet of engineering cars. Specially designed procedures and schedules are used to check individual properties. For example, we have a fade schedule, an effectiveness schedule, and various types of durability schedules. Overall field performance is determined by installing linings on taxicabs, police vehicles, and fleet cars. Cross-country trips to the warm south, to the cold of Minnesota, to the mountains of West Virginia, and to the heights of Pike's Peak are made with test cars. These trips permit first-hand observations of brake performance under various driving conditions. The facilities of the General Motors Proving Grounds at Milford, Michigan and

Mesa, Arizona are used extensively. Test engineers also evaluate linings by driving many miles on city streets and highways under "normal" driving conditions.

Testing of a new lining compound is costly and time consuming. Only a few compositions can be adequately tested in a year's time. This emphasizes the importance of a laboratory test which will separate unsatisfactory materials from promising materials.

Comparison of Metallic with Conventional Linings

Metallic linings have found their greatest acceptance where braking requirements are most severe. However, experience has shown that DM81 linings also have superior qualities in areas other than antifade. They have excellent stability. Effectiveness, though lower than conventional linings, does not change with prolonged use. Lining life is greater than conventional linings under any type of service. The linings are kind to drums since part of the heat is directed away from the drums and into the shoes. Lowering the drum temperature eliminates the tendency for heat checking.

The present metallic linings do not have as high a coefficient as conventional linings. In order to raise the overall output of a brake system it is recommended that a master cylinder with a reduced diameter be used with metal. The greater mechanical advantage afforded by the smaller master cylinder serves to reduce the pedal force and compensates for the lower coefficient. Such a master cylinder requires more pedal travel than the standard master cylinder but, as will be shown, this is not a disadvantage with metallic linings.

The following charts have been prepared to compare performance of metallic linings with conventional linings. In these comparisons the conventional linings were used with 1-inch master cylinders and the metallic linings with $^7/_8$-inch master cylinders. Power brakes were used in all tests.

Fade Resistance

Figure 4 shows the fade resistance of metallic linings as compared with conventional linings. The test consists of making

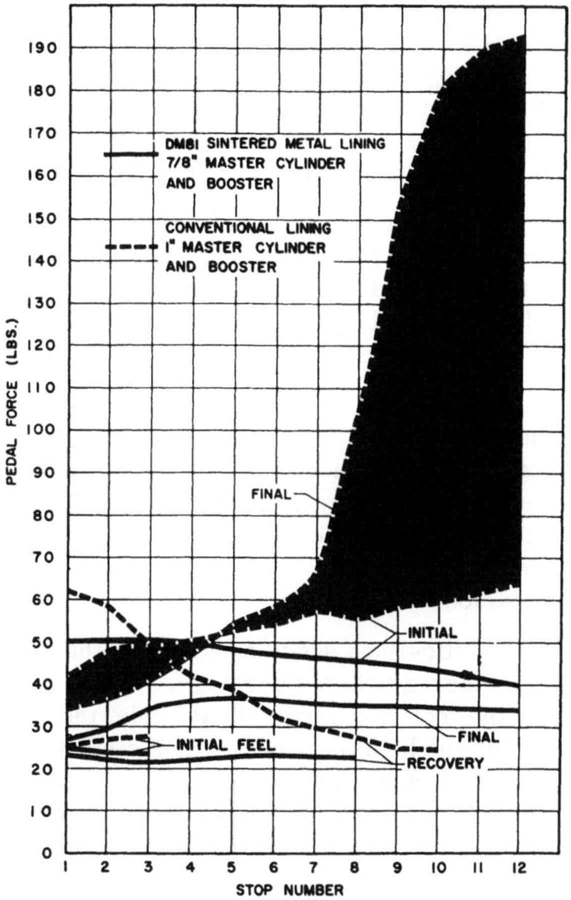

Fig. 4. Fade and recovery stops. Fade: 60 mph, 15 ft/sec^2, 0.4-mile interval, 40-sec interval. Initial feel: 40 mph, 8 ft/sec^2, 0.5-mile interval. Recovery: 40 mph, 8 ft/sec^2, 1.0-mile interval. Pedal travel on 12th stop: conventional 3.50 in., metallic 2.25 in.

12 stops from 60 mph at 0.4 mile and 40 second intervals at a deceleration rate of 15 ft/sec^2. The pedal force required to maintain the 15 ft/sec^2 deceleration at the beginning and end (labeled "initial" and "final") of the stop is recorded. Pedal force in pounds is shown at the left of the chart and stop number at the bottom. Solid lines indicate metallic lining, broken lines indicate conventional lining.

Metallic lining maintained a substantially uniform pedal force for the entire test. Conventional lining required a greater force as the test progressed. At the 12th stop the metal lining still had increasing effectiveness throughout the stop, starting at 40 lb and decreasing to 33 lb. Conventional lining, on the other hand, started at 63 lb and faded, requiring 193 lb at the end.

The curves marked "initial feel" and "recovery" are check stops made from 40 mph at a deceleration rate of 8 ft/sec^2 before and after the fade. Metallic linings required no recovery time since they did not fade, while the conventional linings required nine miles of cooling before they returned to their initial effectiveness.

It should be noted that pedal travel on the final stop with metallic linings was 2¼ inches, while with conventional linings the travel was 3½ inches in spite of the fact that the ⅞-inch master cylinder used with metallic linings inherently requires greater pedal travel for a given line pressure. The lesser travel is possible for two reasons.

1. The metallic-lined shoes expand with the drums as the brakes increase in temperature.
2. Since the metallic linings do not fade, there is no requirement for the high pressures which distort the drums and the entire brake system.

Effectiveness

One of the methods of evaluating brakes is to make stops from various speeds and apply the brakes with a pedal force that increases at a constant rate. A plot is then made of deceleration vs. pedal force: the steeper the curve the greater the lining effectiveness. Figure 5 shows deceleration curves for metal and conventional linings for stops from 70 mph. The two curves follow essentially the same path, indicating that the output of the metallic-lined system is equivalent to that of the conventional-lined system.

Incidentally, these curves were run after the linings had been subjected to 1000 miles of a durability schedule. The conventional lining had greater output when evaluated after break-in, but had fallen to the indicated values with use. Metallic lining effectiveness does not change with service.

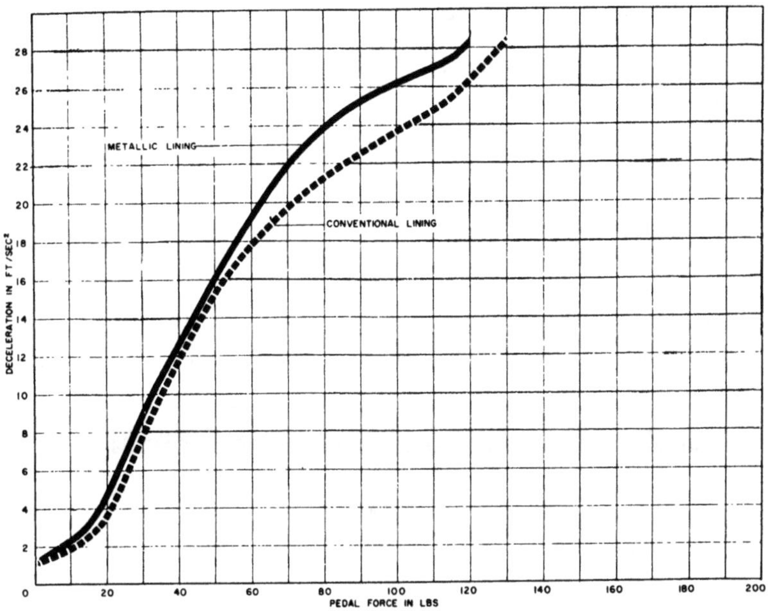

Fig. 5. Brake deceleration from 70 mph.

Wear

Figure 6 gives a comparison of lining life in severe police car duty and in more normal street and highway service. The top half of the chart shows the comparison for severe police duty, the bottom half for normal street and highway service. In severe police work metallic lining averaged 26,000 miles compared to 2000 miles for conventional materials, a ratio of 13 to 1.

The conventional linings were placed at 2000 miles not because they were worn to the shoes, but rather because their effectiveness and stability had been adversely affected by heat and the brakes were unpredictable.

The wear comparison for more normal driving shows 80,500 miles for metallic and 41,000 miles for conventional linings, a ratio of 2 to 1.

Experience has shown that metallic linings wear longer than conventional ones. How much longer depends on the type of service. In general, the more severe the usage the greater will be the difference in life with the advantage in favor of the metallic.

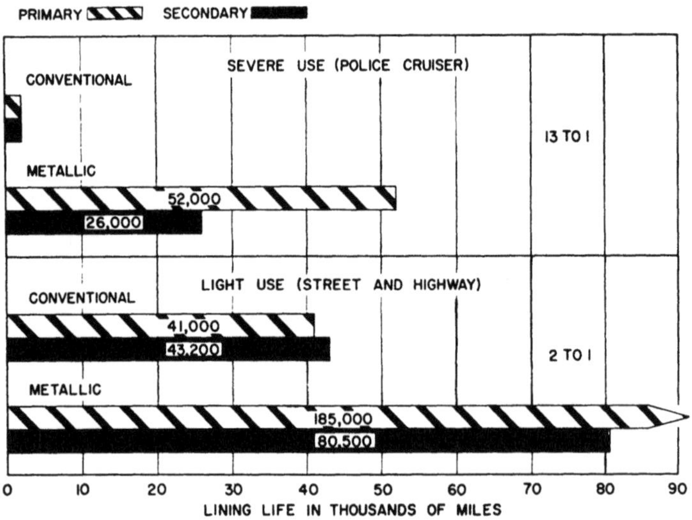

Fig. 6. Lining wear under different conditions of service.

The chart also shows that with conventional linings, in street and highway driving, primary and secondary wear is about equal. With metallic linings the primary wears much more slowly than the secondary. This difference in wear rate is used to advantage by making a thick–thin lining combination. The primary lining is made thinner and the secondary thicker than standard.

One of the most important characteristics of any lining is its stability under all conditions. Slight leads and pulls due to unequal output of front brakes are aggravating to the driver. Severe pulls can be dangerous.

In this area of stability, metallic linings have an excellent record. During the many, many thousands of road miles on DM81 material, complaints of brake pull have been virtually nonexistent. This good stability is due primarily to the fact that metallic linings are sintered at temperatures above that encountered in brake use. Thus, no change can take place in their composition or friction regardless of the type of usage. A further advantage of the linings is that they are relatively insensitive to contamination by water and grease.

Conclusion

While much has been accomplished in the field of metallic linings, we realize that we still do not have all the answers. For this reason a continuous development and testing program is in progress at our plant.

The successful application of sintered metal friction material for automotive brake linings opens a new field for the metal powder industry.

Let us take a brave look into the future. A car set of metallic linings for a standard size car requires 68 segments weighing a total of 7.3 lb. Suppose one of the "big three" should produce 1,000,000 cars in a year and decided to adopt this lining across the board. As many as 68 million pieces and 7 million pounds of powder would be required! Think of the number of presses, furnaces, and other allied equipment needed for such production! Going a step further, suppose all cars should adopt metallic lining. And we haven't touched the service business. Neither have we considered trucks and buses.

Thus, all of us in the industry – suppliers of raw material, fabricators, and manufacturers of equipment – can look to this one single application of metal powder with anticipation and promise. The problems are many and complex, but the rewards are great.

Chapter 3

A Contribution to the Investigation of Dry Friction of Sintered Steel

V. Procházka and V. Miskovic

Laboratory of Metallurgical Technology
Technical University
Košice, Czechoslovakia

When two dry surfaces are rubbed against one another, a mutual penetration always takes place due to the roughness of the surfaces and variations in their structures. The subsequent deterioration and damage of the rubbing surface is caused by a complex variety of processes, which can be classified as microcutting or grooving, fatigue or checking, and "deep plucking" or shell-out.

Shell-out is a most undesirable phenomenon. It occurs during the sliding of two bodies, when the separation of welded spots takes place at a certain distance from the interface in one of the bodies. The surfaces of both bodies rapidly become extremely rough and display grooves of various lengths. The bottom and edges of these grooves are very rough (in contrast to grooves caused by microcutting). Shell-out is also accompanied by intense changes of the surface layers to a considerable depth. However complicated a phenomenon it may be, it is primarily caused by seizure due to interatomic adhesion forces, which become active wherever a true contact of metallic surfaces occurs.

A number of theories have been advanced in an attempt to explain the adhesion of two metallic surfaces as a result of recrystallization, diffusion, change in the state of energy, and also an interaction between the metals and their oxide layers.

A theory based on recrystallization was first proposed by Parks [1]. It is based on the following assumptions. Individual grains of rubbing surfaces are differently oriented and deformed. Because of the increased temperature recrystallization takes place on the surfaces, which may result in the seizure of grains that had been oriented differently. Oxide surface films also coalesce during the recrystallization, thus removing another obstacle to metallic adhesion.

Lashko [2] proposes a diffusion theory which describes seizure as a diffusion process.

Another theory has been worked out by Tylecote [3], who considers the mechanical properties of oxide films and base metals. The ratio between the hardness of the oxide film and that of the metal decisively affects the deformation necessary for the formation of a junction.

An important theory, known as energetic or diffusionless, has been offered by Semenov [4] who claims that the mere proximity of metallic surfaces to the point where atomic attraction forces become active is not sufficient to cause seizure. Because of the different lattice orientation of the surfaces a definite displacement of outer atoms from their equilibrium position is necessary before a junction can be formed. Therefore, to form a junction without diffusion or recrystallization requires a definite activation energy, which can be supplied either by heat or deformation. Erdmann-Jesnitzer also supports the diffusionless theory on the basis of his own experimental results [5, 6].

Our work covers the study of seizure which occurs upon friction of P/M iron–graphite against cast iron. The dry friction of iron–graphite is being studied in several countries, but mainly in the USSR and East Germany, as indicated by the papers of Bebnew [10, 11], Rakovskij [12], Nardov [13], Strobel [14], Eisenkolb [15], etc. The tendency of seizure between different structures of carbon steel and a slider made from wear-resistant chromium steel had been previously investigated by Navara [8]. His results, obtained by measuring the amplitude of stick-slip motion under a light load and a small frictional velocity, indicated that fine pearlite was very resistant to adhesion, whereas ferrite is not.

Selection, Preparation, and Testing of Specimens

Graphite is universally used to improve sliding properties of metals. Its lubricating ability is explained by the hexagonal structure of its crystal lattice, where individual planes can easily slide on each other [7]. Furthermore, graphite adheres strongly to steel. As a result, shearing between rubbing bodies occurs either within the graphite itself, or between graphite and the metal. Metal-to-metal contact is almost eliminated.

There are a number of grades of cast iron containing from 2.4 to 4.2% carbon and 1.3 to 2.4% silicon [7] in which uniformly dispersed graphite greatly improves the frictional properties.

Powder metallurgy makes it possible to produce compacts with any desired amount of graphite. However, for practical application only enough graphite should be added to assure sufficiently strong products without requiring complicated manufacturing technologies, such as hot pressing.

Preliminary experiments have shown that specimens with up to 26% by weight of graphite possess adequate strength properties. Further experiments were therefore made with 4 to 26% of graphite in increments of 4%.

The best specimens were made from a mixture of iron powder below 250 mesh and a natural graphite, also below 250 mesh. The graphite's carbon content was 88%. A mixture of iron and graphite powders was blended for 4 hr in a laboratory drum blender. 10 × 10 mm square specimens were compacted at a specific pressure 5 Mp/cm^2 and were sintered in a charcoal pack and hydrogen atmosphere at temperatures between 1050 and 1100°C.

Frictional tests were performed on an SFM-1 test apparatus, which simulates conditions prevailing in normal operation, i.e., gradual rise in temperature and formation of oxide film. Two specimens are clamped in holders on the opposite ends of a steel segment which could turn around its axis. A rotating cast iron disc is pressed against the two specimens; the velocity of rotation and the applied load were adjustable. The torsional moment which

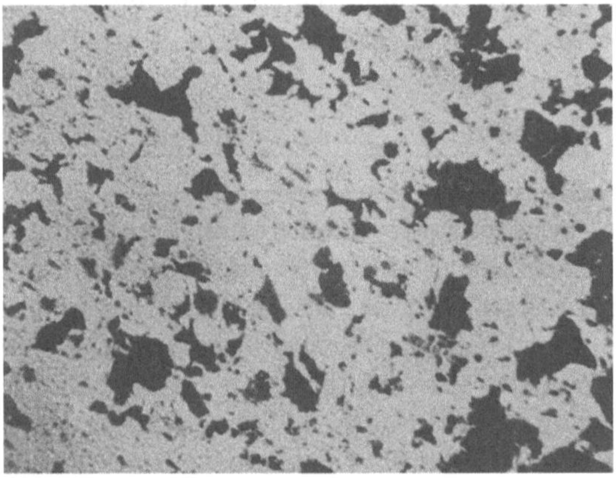

Fig. 1. Micrograph of specimen with 8% added graphite. Unetched (80×).

carries the segment is counterbalanced by a pendulum attached firmly to the segment. The displacement of the pendulum, proportional to the coefficient of friction, is measured by an induction scanner; a core connected to the pendulum is pushed into an induction coil, the inductance of which is being measured and registered. Temperature is measured by a thermocouple, the end of which is inserted in a hole drilled into the specimen 1 mm from the rubbing surface. A typical test lasts one hour, which ensures stabilization of all parameters. Wear is measured by weighing the specimens before and after each test and is evaluated as specific wear in grams per hour per horsepower.

The Effect of Graphite Addition on Frictional Properties

Specimens containing 4, 8, 12, 16, 24, and 26 wt.% graphite were also tested at a sliding velocity of 10 m/sec and a specific load of 9 kp/cm^2. The metallic structure consisted of fine pearlite, which had formerly exhibited very good resistance to seizure [8]. Micrographs of unetched specimens containing 8 and 24% graphite are shown in Figs. 1 and 2.

Test results show that the amount of added graphite has a significant effect on the frictional properties, especially on the wear of the specimens. A pronounced decrease in wear occurs

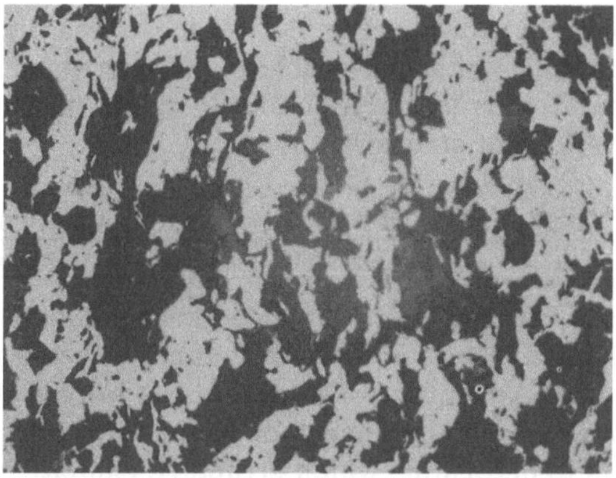

Fig. 2. Micrograph of specimen with 24% added graphite. Unetched (80×).

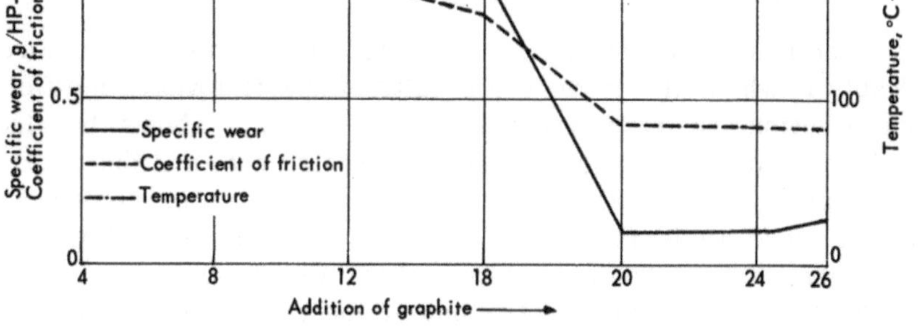

Fig. 3. Coefficient of friction, specific wear, and temperature as functions of added graphite. Load 9 kp/cm^2, sliding velocity 10 m/sec.

when the amount of added graphite exceeds 16%. The dependence of the coefficient of friction, wear, and temperature on the amount of added graphite is shown in Fig. 3. All measured properties decrease with up to 20% of added graphite. When graphite exceeds about 24%, wear begins to increase slowly again, presumably due to an excessive loss of strength.

The Effect of Distribution and Amount of Cementite upon Frictional Properties

Specimens containing 24% of added graphite and sintered at 1075°C were heat-treated to obtain structures with varying amounts and distribution of cementite. The specimens were tested on the SFM-1 apparatus at a sliding speed of 10 m/sec. The purpose of these tests was to determine which structure exhibited the best frictional properties, and also whether the results of previous experiments, performed under light loads and at low velocities [8], were applicable to conditions of increased temperature where oxidation took place. Specimens with the following metallographic characteristics were prepared:

a. Pearlite with cementite network along grain boundaries (Fig. 4).
b. Coarse lamellar pearlite (Fig. 5).
c. Fine pearlite (Fig. 6).
d. Globular pearlite (Fig. 7).
e. Pearlite with ferrite (Fig. 8).
f. Ferrite with pearlite (Fig. 9).

The heat treatment of the individual specimens is described in Table 1, which also gives the results of specific wear, coefficient of friction and friction temperature. The table shows that fine pearlite demonstrates the highest wear resistance, and that globular pearlite differs little in the tested properties. Coarsening of lamellar pearlite decreases the friction properties, though not very significantly. The appearance of ferrite in the structure is very harmful and free cementite has an intermediate negative effect. Figure 10 illustrates the frictional behavior of the test structures.

The friction tests performed have proved that the results of previous investigations [8] can be applied to practical operating conditions.

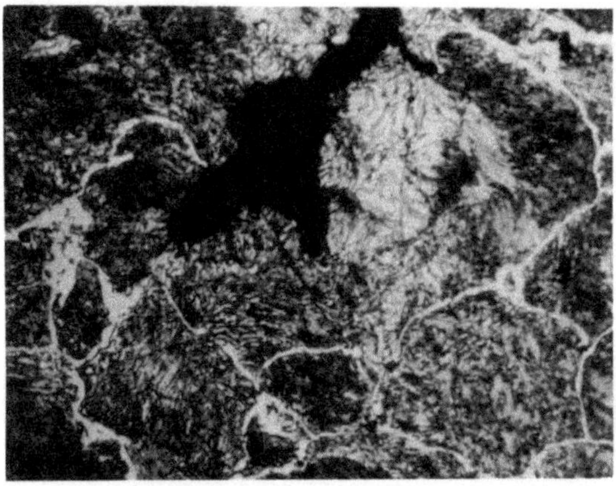

Fig. 4. Micrograph of specimen of fine pearlite with cementitic network. Nital etch (500×).

Fig. 5. Micrograph of specimen of coarse lamellar pearlite. Nital etch (500×).

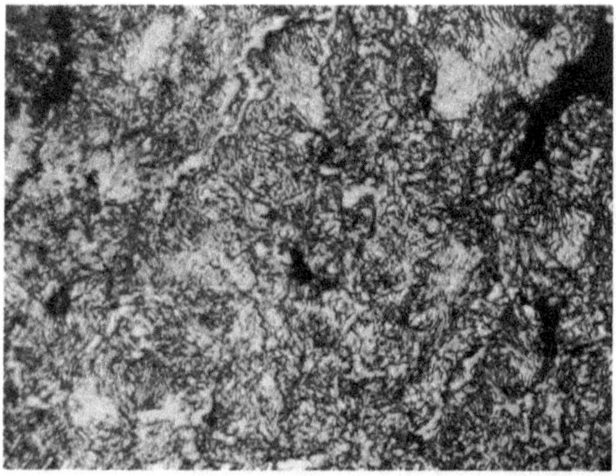

Fig. 6. Micrograph of specimen of fine pearlite. Nital etch (500×).

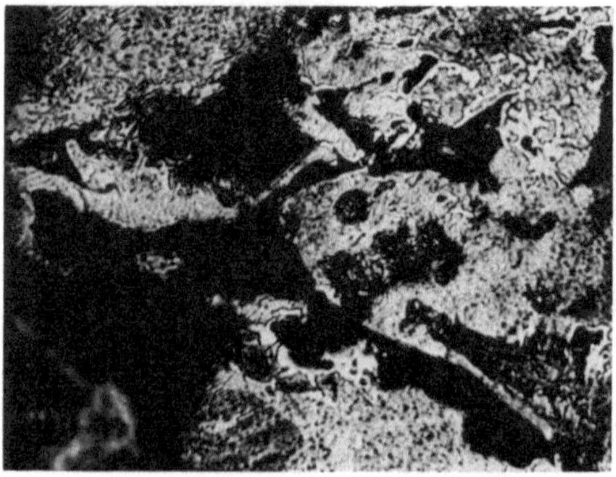

Fig. 7. Micrograph of specimen of globular pearlite. Nital etch (500×).

Fig. 8. Micrograph of specimen of pearlite and ferrite. Nital etch (500×).

Fig. 9. Micrograph of specimen of ferrite and pearlite. Nital etch (500×).

TABLE 1. Frictional Characteristics of Various Types of Structures at a Sliding Velocity of 10 m/sec

Type of structure	Specific wear, g/HP·hr	Coefficient of friction	Temperature, °C	Heat treatment
Lamellar pearlite with cementitic network along grain boundaries	0.29	0.52	239	Structure stabilized by addition of 0.4% phosphorus; free cooling from the sintering temperature
Coarse lamellar pearlite	0.19	0.48	225	Free cooling in the retort of the furnace.
Fine lamellar pearlite	0.1	0.42	211	Intensified cooling by a stream of air
Fine globular pearlite	0.14	0.44	220	Isothermal decomposition for 1 hr at 720°C
Pearlite + ferrite	0.40	0.55	246	Isothermal decomposition for 5 hr at 720°C
Ferrite + pearlite	0.54	0.61	272	Isothermal decomposition for 14 hr at 720°C

The Effect of Sliding Velocity on Seizure

Specimens with 24% graphite added, and possessing a fine pearlitic structure have been tested at sliding velocities ranging from 0.7 to 13.5 m/sec: the results are plotted in Fig. 11. Both wear and the coefficient of friction change significantly with varying velocity. The specimens are wear-resistant at low and high speeds, while in the intermediate ranges their wear is rather high. The dependence of the coefficient of friction on speed is similar. The friction temperature increases regularly with increasing velocity.

Most remarkable is the steep rise in wear and friction coefficient when the velocity increases from 0.7 to 1.4 m/sec. Both

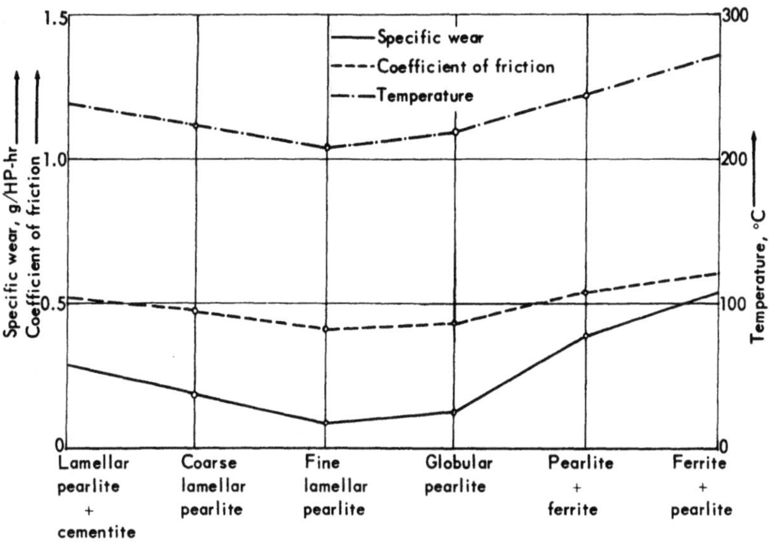

Fig. 10. Frictional characteristics as a function of microstructure. Load 9 kp/cm^2, sliding velocity 10 m/sec.

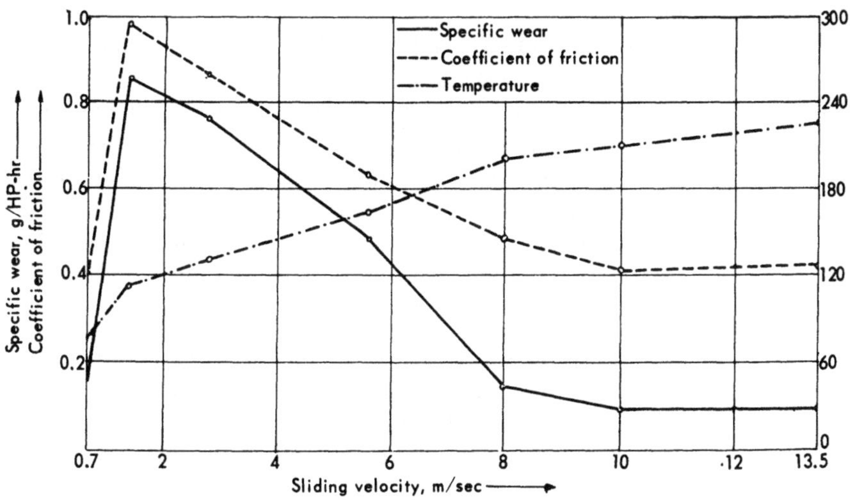

Fig. 11. Frictional characteristics as a function of sliding velocity. Load 9 kp/cm^2, fine pearlitic structure.

parameters have their highest values at the latter sliding speed. With a further increase in velocity, wear and the coefficient of friction gradually fall again and become roughly constant at higher velocities.

The explanation is presumably the temperature of friction. Under low velocities when the temperature does not exceed 100°C, graphite acts as a reliable lubricant and separates the two metallic surfaces by a thin film. Above 100°C, the lubricating ability of graphite diminishes rapidly and the subsequent contact of rubbing surfaces causes intense seizure and excessive wear. And the temperature is still too low to cause a rapid formation of oxide films, which would obstruct the seizure phenomenon. Only on a further increase in sliding velocity and temperature will the oxide films prevent metallic bonding, and wear and a decrease in the friction coefficient.

To complete the picture of seizure, tests of different iron phase structures were repeated at the velocity of 1.4 m/sec. Four types of structures, shown in Figs. 4, 6, 8, and 9, were tested. Fine pearlite again proved to be the most resistant to seizure. The presence of ferrite deteriorated frictional properties, just as in the case of sliding at 10 m/sec. The dependence of wear, coefficient of friction, and temperature on the type of structure at a sliding speed of 1.4 m/sec is shown in Fig. 12. The action of the graphite film, ensuring low coefficient of friction and low wear under normal conditions, i.e., in air and at room temperature, is well known in electric brush applications [9]. It has been established, however, that the lamellar structure of graphite is, in itself, insufficient to ensure lubrication, because the adsorption ability of surface atoms is the cause of low friction and wear. If the conditions change, especially if the partial pressure of water vapor falls below about 3 mm Hg, the adsorbed water evaporates and the lubricating ability of graphite disappears.

A similar situation is the lack of adsorbed water molecules which occurs when the temperature increases above 100°C. This probably can account for the deterioration of the lubricating ability of graphite under certain of the above-described conditions.

Conclusion

The results of studying the frictional properties of P/M iron–graphite show that the formation of metallic junctions de-

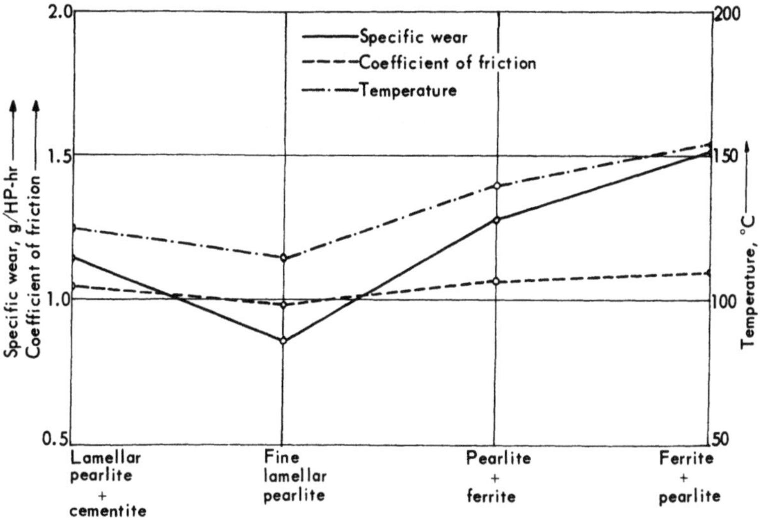

Fig. 12. Frictional characteristics as a function of microstructure. Load 9 kp/cm^2, sliding velocity 1.4 m/sec.

pends on the nature and amount of cementite in the structure, on the intensity of surface oxidation, and on the amount of added graphite.

Testing different structural components proved that a pearlitic grain, consisting of a fine pattern of ordered crystallites of ferrite and cementite, is very resistant to seizure. When ferritic grains appear in the structure the frictional properties deteriorate due to the increased tendency of seizure between the grains of equal crystallographic structure.

It has further been proved that a definite amount of graphite is needed to restrict the phenomenon of seizure between the metallic components. An increase in sliding velocity leading to a rise in temperature, lowers the wear, at least in a significant range of temperatures. This occurs because of the formation of an oxide film which obstructs direct metallic contact and the formation of metallic junctions.

Seizure of metallic components is most apparent in a temperature range where no recrystallization or intense surface oxidation occurs. The diffusion processes cannot affect the phenomenon of seizure at relatively low temperatures.

After considering the various theories dealing with the problems and causes of seizure and examining the described experimental results, it can be concluded that the probable cause of seizure is in the correlation of crystal lattices and the energy fluctuations supported by pressure at the points of true metallic contact.

References

1. Parks, E., The Welding Journal, Vol. 32, No. 5, 1953.
2. Lashko, N. F., and Lashko-Avakyan, S. V., Metallvedenie svarki, Mashgiz, 1954.
3. Tylecote, R. F., Metal Industry, Vol. 81, No. 4, 1952.
4. Semenov, A. P., The Seizure of Metals, Mashgiz, 1958.
5. Erdmann-Jesnitzer, F., Zeitschrift Metallkunde, No. 12, p. 854, 1955.
6. Erdmann-Jesnitzer, F., Zeitschrift Metallkunde, No. 12, p. 756, 1955.
7. Petrichenko, V. K., Antifriction Materials and Plain Bearings, GNTIML, Moscow, 1954.
8. Navara, E., Problems of Powder Metallurgy, VSAV, Bratislava, p. 127, 1964.
9. Shmayler, A., Progress in Powder Metallurgy, CSAV, Prague, p. 313, 1954.
10. Bebnev, P. I., Powder Metallurgy, Yaroslavl, p. 223, 1956.
11. Bebnev, P. I., Powder Metallurgy, Yaroslavl, p. 90, 1956.
12. Rakovskij, V. S., Fundaments of Physical Powder Metallurgy, GNTI, Moscow, 1962.
13. Nardov, K. M., Properties of Friction Materials, Izd. Akad. Nauk SSSR, Moscow, 1959, p. 88.
14. Strobel, E., Proceedings of the 2nd International Powder Metallurgy Conference in Eisenach, 1961, p. 541.
15. Eisenkolb, F., Proceedings of the 2nd International Powder Metallurgy Conference in Eisenach, 1961, p. 61.

Chapter 4

The Influence of Structure and Metal Additions on the Properties of Iron-Graphite Sintered Materials

E. Strobel and H. Rebsch

Institute for Special Materials
Dresden, DDR.

Introduction

The use of iron–graphite P/M materials for bearing and friction applications has been increasing in recent years. They have proved successful and are beginning to gain over materials based on the copper–graphite system. New American [1] and English [2] papers report that a considerable amount of iron-base friction material is already being produced. Several papers have appeared in Czechoslovakia [4-7] on the application of iron–graphite for frictional purposes. It seems that even Japan has started the production of sintered iron–graphite in large quantities [3]. In Germany iron–graphite sintered materials were developed a few years ago and successfully introduced into practice [9-14]. Bearing materials on the same basis have also proved successful and found industrial application.

Iron–graphite sintered materials contain, beside their main components, other metallic and often nonmetallic additions which

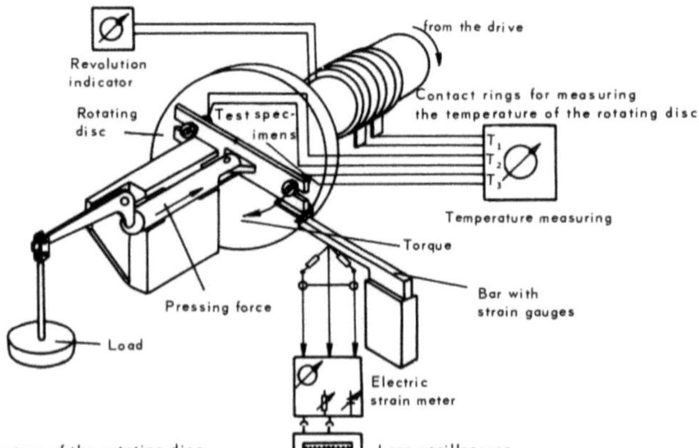

Fig. 1. Friction test apparatus.

contribute to strength and frictional properties. However, these properties depend to a considerable extent upon the microstructure of the iron component, which can be influenced by alloying additions and thermal treatment. Before referring to these problems in detail, the friction test apparatus used for the experiments will be described.

Friction Test Apparatus

The test apparatus used to evaluate frictional properties is shown in Fig. 1. Two specimens having a surface area of 6.3 cm^2 are pressed by means of a mass load against a rotating disc. The rotation moment due to friction between the specimens and the disc is supported by a rigid but flexible bar on which strain gauges are mounted. The deflection of the bar is measured by means of an electric strain meter, and from the measured values the torque and also the coefficient of friction can be calculated. This device can be coupled with a loop oscilloscope to register the variation of the coefficient of friction. The temperatures of the specimens and of the rotating disc are measured by special thermoelements. The current resulting from the thermoelectric force is conducted through silver–graphite contacts sliding on silver-coated rings. The wear of the specimens is measured in terms of decrease of their thickness and evaluated as the specific wear related to one hour's duration of 1 Hp output. Individual frictional tests are performed for 45 minutes. The equilibrium in temperature is usually attained in 5 to 15 minutes, depending on the frictional power.

The Influence of Cooling Rate on the Structure and Properties of Iron – Graphite Sintered Material

Sintered iron–graphite with a high carbon content behaves differently from sintered steel with low carbon content during both sintering and cooling. A high proportion of free graphite acts against the formation of cementite [13, 15]. If the addition of graphite exceeds 6 to 8 wt.%, there is usually no secondary cementite in the structure; ferrite is often present in addition to pearlite. This phenomenon has been explained by the fact that the numerous fine graphite particles present act as nuclei for graphite precipitation, which prevents the formation of cementite. The structure can be affected by cooling after sintering and/or by another subsequent thermal treatment. However, the resulting structure will determine the frictional properties of sintered iron–graphite only if it does not subsequently change due to temperatures evolved from the process of friction. And it seldom happens that the frictional power and the evolved temperature are so low that no change in structure takes place. Higher powers and the resulting temperatures rapidly change the structure of the rubbing surface, so that its influence is minimal or disappears completely. This is especially true when the temperature of the rubbing surface exceeds the gamma region. Further investigation is needed to determine how the strength of the bulk specimen will be affected. The highest temperatures appear during friction on the surface layers of the rubbing materials. Due to the good heat conductivity of sintered iron–graphite, the temperature drop between the surface and the bulk is steep, and substantial changes in the structure of the bulk material do not occur unless the frictional power is very high and the process of rubbing lasts for a long period.

Stresses similar to those imposed on the friction element apply to the counterface. It is therefore essential that, when sintered friction elements are used, the material of the counterbody is carefully selected so that it does not change its structure or mechanical properties during the rubbing process. Only by paying close attention to these rules is it possible to take full advantage of metallic friction materials [14].

Figure 2 shows the microstructure of sintered iron–graphite with 17.5 wt.% graphite, and the effect of the cooling rate. The specimens were sintered at 1050°C in an inert atmosphere and cooled at different rates from this temperature. The pearlitic

Fig. 2. Structure of sintered iron–graphite as affected by different cooling rates (graphite content 17.5%, sintering temperature 1050°C, enlarged 500:1, etched in 2% alcoholic nitric acid).

structure is finer with a faster cooling rate, until finally martensite appears. The micrographs also show small areas of ferrite in addition to the prevalent pearlite. However, secondary cementite can hardly be noticed.

Figure 3 gives the mechanical properties and frictional behavior of the tested sintered iron–graphite. The strength properties of slowly cooled material are substantially lower, and increase with a rising rate of cooling. Hardness and compressive strength increase steadily, whereas bending strength decreases again when martensite appears. Both coefficient of friction and wear increase markedly, at a sliding velocity of 6 m/sec, with decreasing cooling rate. This tendency is visible even at the sliding speed of 21 m/sec, but the slope of the curves is substantially flatter. The influence of the structure of iron–graphite is significant only at very low sliding speeds and temperatures. In this case, the wear decreases as the fineness and hardness increases, while the coefficient of friction increases as the structure becomes coarser.

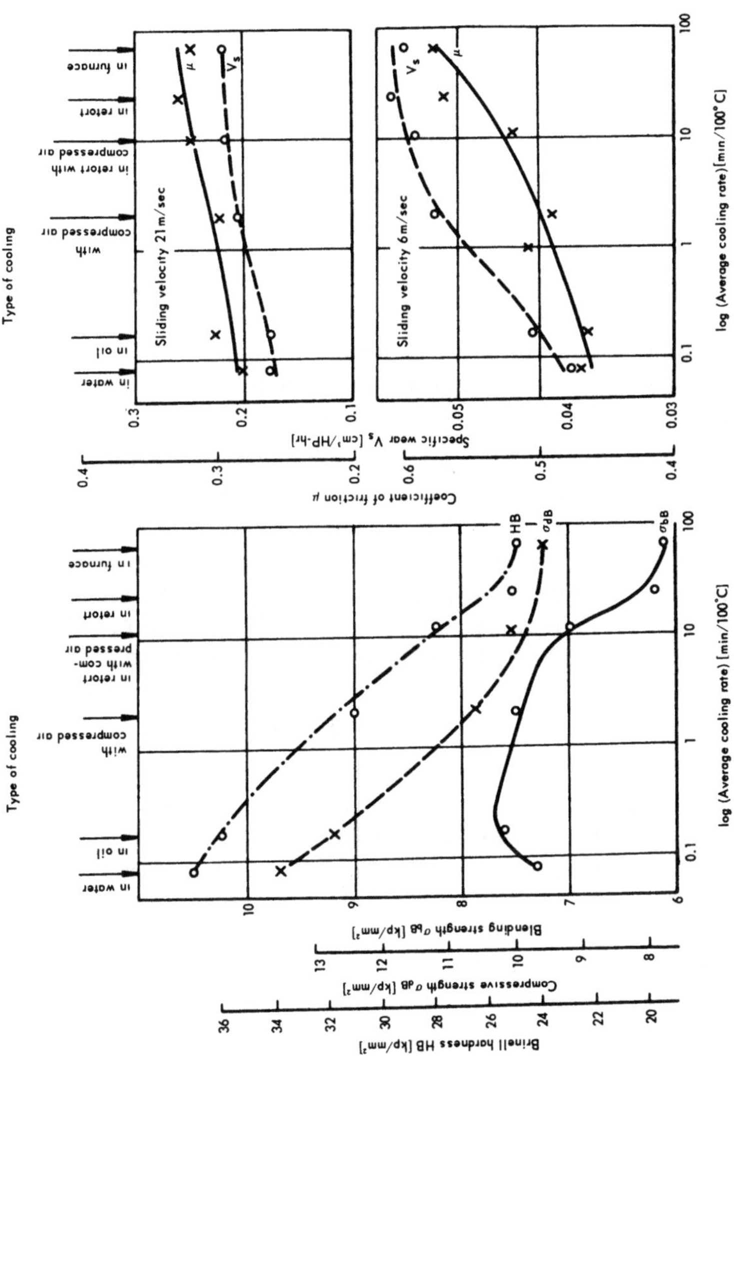

Fig. 3. Mechanical properties and friction characteristics of sintered iron–graphite as a function of the rate of cooling.

The Effect of Copper Additions on the Structure and Properties of Sintered Iron – Graphite

The structure of sintered iron–graphite can be affected by the addition of alloying components. Copper was the first element tested, as it had often been recommended for the improvement of sintered iron–graphite [16-18]. Copper additions to sintered steel have often been used because they activate the sintering process by forming a liquid phase and thus increase the strength. Also, the compensation of shrinkage attainable by copper has received much attention [19]. These advantages which copper offers apply to sintered iron–graphite with a high amount of graphite in a limited manner only. The liquid phase formed during sintering does not wet graphite which is present in larger quantities, so that its positive effect on the sintering process is minor. As iron–graphite already shows some volume increase during sintering, additional growth caused by copper would be a disadvantage.

An investigation of the effect of copper on the properties of iron–graphite has been carried out by L. Schneider of the Institute of Materials Science of the Technical University in Dresden. Copper powder was added to a mixture of iron and graphite. In order to keep the amount of graphite constant, the amount of iron was lowered in all cases by the amount of added copper. The specimens were sintered at 1120°C in an inert atmosphere. This temperature makes possible the formation of the liquid phase and ensures the complete dissolution of copper.

Figure 4 shows the microstructure of sintered iron–graphite as a function of the content of copper. The structure without copper consists of pearlite with very little ferrite. It should be noted that as the amount of copper increases, more ferrite is formed at the expense of pearlite. When the amount of copper reaches 5%, free crystals of copper-rich alloy appear, especially when contact with iron is obstructed by graphite. The addition of copper therefore supports the precipitation of graphite in sintered iron–graphite. Finally, the structure of the material with 10% copper consists almost exclusively of ferrite and copper-rich solid solution. Similar results have been reached by A. Domsa and coworkers [20, 21], who have investigated the influence of copper additions to iron–graphite bearing materials with 4% graphite. They consider it very advantageous that copper suppresses the

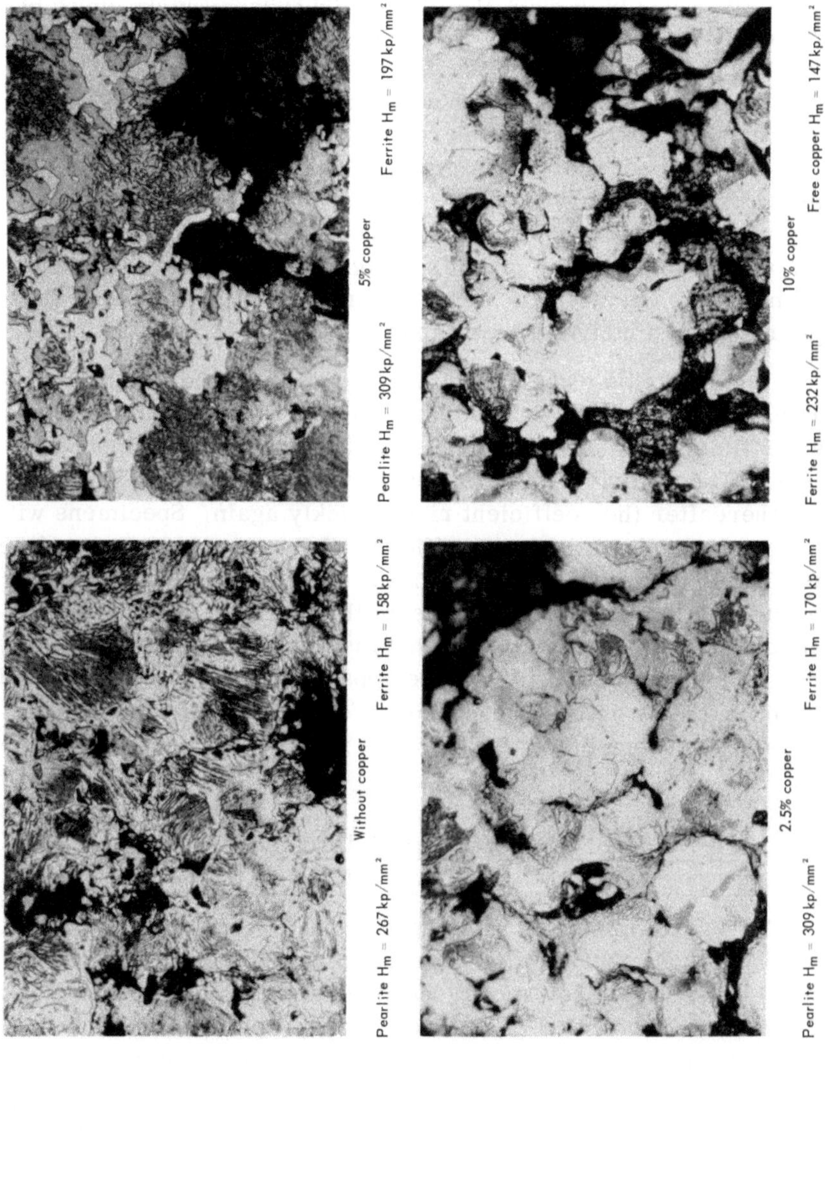

Fig. 4. Structure of sintered iron–graphite with copper addition (graphite content 17.5%, sintering temperature 1120°C, enlarged 250:1, etched in 3% alcoholic nitric acid).

formation of free cementite, which is harmful in bearing materials. This is, however, of no significance in the case of sintered iron–graphite with a high content of graphite, as very little cementite is formed in this case. The data on the microhardness of the structural phases given in Fig. 4 show that copper increases the hardness of both pearlite and ferrite. However, ferrite is in all cases less hard than copper-free pearlite. Considering the increasing portion of ferrite caused by copper additions, it is easy to predict that the strength properties decrease with added copper (see Fig. 5). It has already been stated that graphite is not wetted by the liquid phase during sintering, and therefore copper cannot substantially improve the sintering process. It is possible to increase the hardness of an iron–copper–graphite system by precipitation hardening; however, the attained values are in every case far below those of martensitic sintered iron–graphite.

Figure 5 also shows the effect of copper on the frictional properties. While wear is only slightly influenced by copper additions, the coefficient of friction decreases at first with small amounts of copper. The minimum value is reached between 2.5 and 5% of copper; thereafter the coefficient rises quickly again. Specimens with 10% copper display a higher coefficient of friction than copper-free material. It is probable that the initial increase in hardness of pearlite and ferrite is responsible for the decrease of the coefficient of friction. Its subsequent increase is connected with the preferential forming of ferrite and the appearance of copper-rich solid solution in the structure. It is therefore assumed that these phases are responsible for the increase in the coefficient of friction.

The Effect of Nickel Additions on the Properties of Sintered Iron – Graphite

Nickel, an important alloying element in sintered steel, increases strength and ductility considerably [22, 23]. It is interesting to investigate how nickel additions affect the mechanical and frictional properties of sintered iron–graphite with a high graphite content. For this purpose, nickel powder [250 mesh] was admixed in various proportions to a mixture of iron and 17.5% graphite. The specimens were compacted at 5 Mp/cm^2 and sintered at 1050°C for 3 hours in an inert atmosphere. Mechanical properties are shown in Fig. 6. Hardness, compressive strength, and

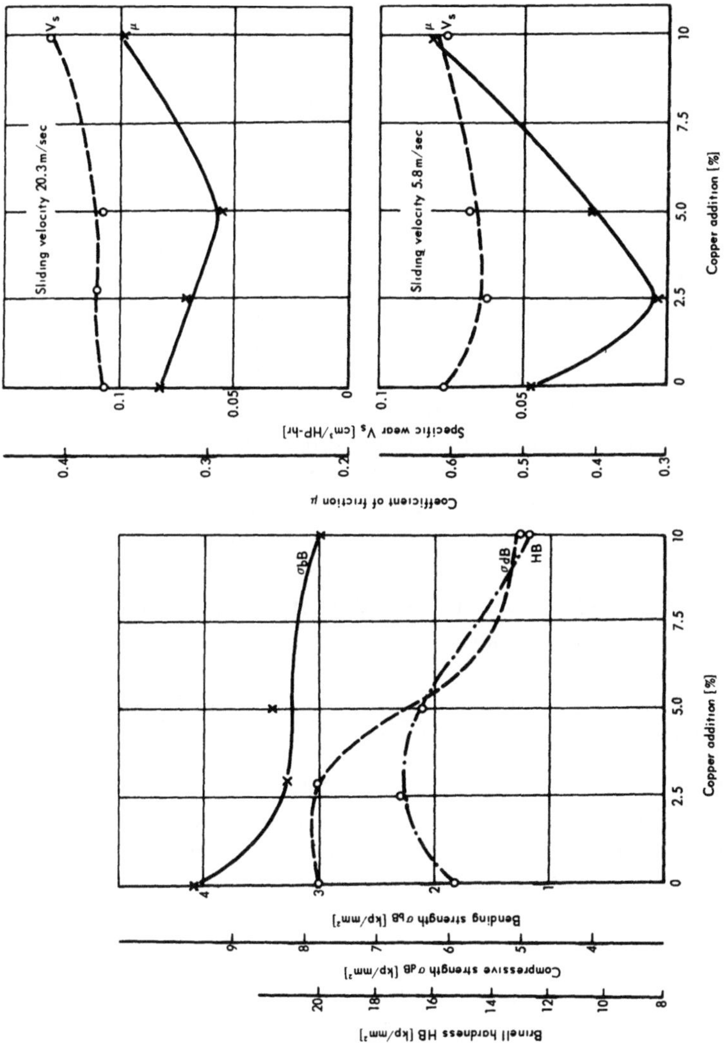

Fig. 5. Mechanical properties and friction characteristics of sintered iron–graphite with copper addition.

bending strength rapidly increase with growing proportion of nickel up to 12.5%, just as it has been observed with sintered steel without or with small amounts of carbon. Metallographic investigations showed that the purely pearlitic structure of sintered iron–graphite had changed due to nickel additions into a pearlitic-ferritic-martensitic structure. Microhardness measurements proved that the increase in the hardness of the pearlitic and ferritic phases caused by nickel was insignificant. When the content of nickel exceeds 2.5%, martensite appears in the structure even at very low cooling rates. This phase displays a microhardness of H_m = 850 kp/mm^2, which is 2.5 to 2.8 times more than the value for pearlite. This is the source of the increased mechanical properties.

Figure 6 also illustrates the frictional properties of sintered iron–graphite alloyed with nickel. It can be seen that at low velocities of sliding, i.e., low friction powers, nickel improves frictional properties; the coefficient of friction increases a little and wear decreases. At higher sliding speeds, on the other hand, the change of frictional properties caused by nickel is almost negligible. It can be assumed that under these frictional conditions high temperatures develop on the rubbing surface, which leads to a corresponding phase transformation. As austenite is most probably being formed, it seems that its frictional properties are independent of nickel content. Further investigations of this problem should be carried out.

The Effect of Low-Melting Metals on the Properties of Sintered Iron – Graphite

Friction and bearing materials often contain, beside graphite and dry lubricants, additions of soft metals, which are expected to form additional lubricating films at high temperatures. It is supposed that these metals, when their melting point or melting interval is reached, sweat out of the surface just as oil does in porous bearings and thus contribute to surface-film lubrication.

Experiments were carried out in order to check the effect of these metals. A selection of nonferrous metals with different melting points was added in powder form to the mixture of iron and graphite. The mixtures were finally compacted in the same manner as the earlier ones and sintered at 1020°C in an inert atmosphere.

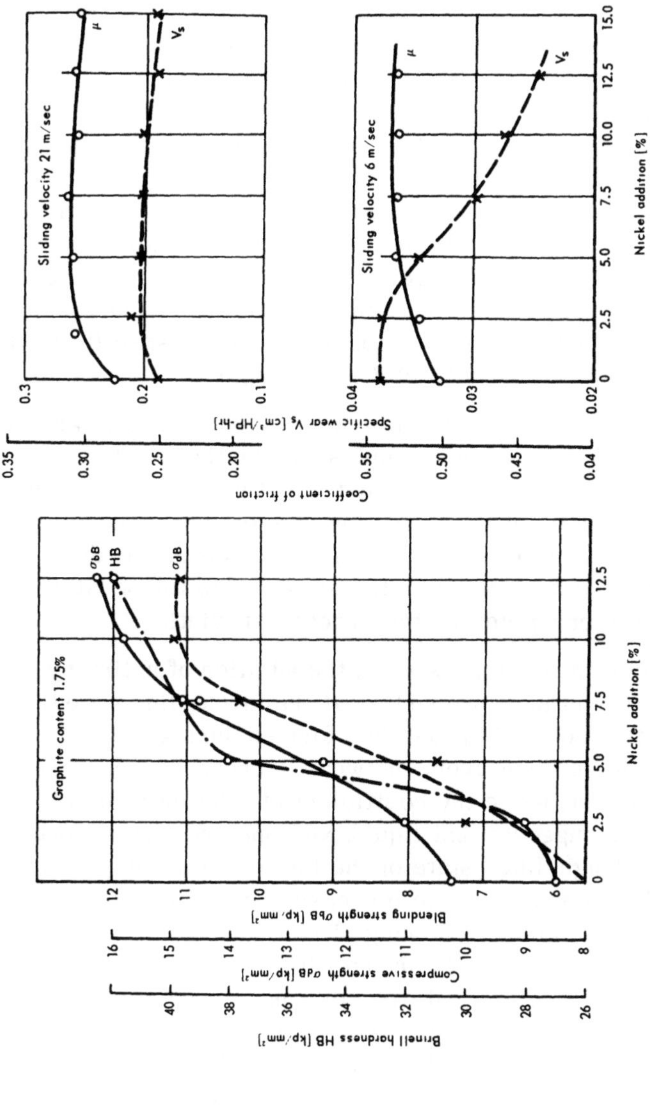

Fig. 6. Mechanical properties and friction characteristics of sintered iron–graphite with nickel addition.

As the highest admixed quantities of low-melting metals did not exceed 10%, traces of sweating were very seldom observed on the surface of the specimens.

Figure 7 shows some characteristic results of these experiments, including the results of the following tested additions: antimony, which melts at 630°C; lead, which melts at 327°C; and a eutectic alloy of 25% tin, 25% lead, and 50% bismuth, which melts at only 94°C. As to the question of the extent to which these metals form solid solutions with iron resulting in liquefaction at different temperatures, it can be answered only with regard to lead or bismuth, since these metals do not alloy with solid or melted iron. However, the results point to an observable dependence of the friction properties on the melting point of the added metal. The frictional behavior was tested at a constant specific load of 6 kp/cm^2 and sliding speeds of 6 and 21 m/sec.

Figure 7 shows the coefficients of friction and wear, and the temperature of the specimens measured by thermoelements. These measurements were carried out at a distance 1.5 mm from the rubbing surface at the beginning of each test. Therefore the actual temperature of the rubbing surfaces cannot be determined from the given data. However the slope of the temperature curve can be used to characterize the process of friction.

As curve a in Fig. 7 shows, the addition of antimony causes no change in frictional properties at the low sliding speed, thus intimating that no liquid phase has formed during the test. At a speed of 21 m/sec, however, the wear decreases with increasing antimony content, while the coefficient of friction rises slightly. This behavior suggests that liquid portions are already present; possibly a "dough-like" state of the low-melting component is attained, which brings about an increase in the coefficent of friction. The formation of an easy flowing phase causes a decrease in the coefficient of friction, as is shown with the addition of lead. In this case the coefficient and wear decrease even at small frictional powers. At the same time the temperature decreases due to additional lubrication. At a higher sliding speed, this phenomenon is understandably more pronounced. Similar results with lead were reached with the eutectic tin–lead–bismuth alloy, where the liquid phase formed even sooner because of the lower melting temperature. The result is a sufficient amount of "lubricating liquid" forming very rapidly, which causes an immediate sharp decrease

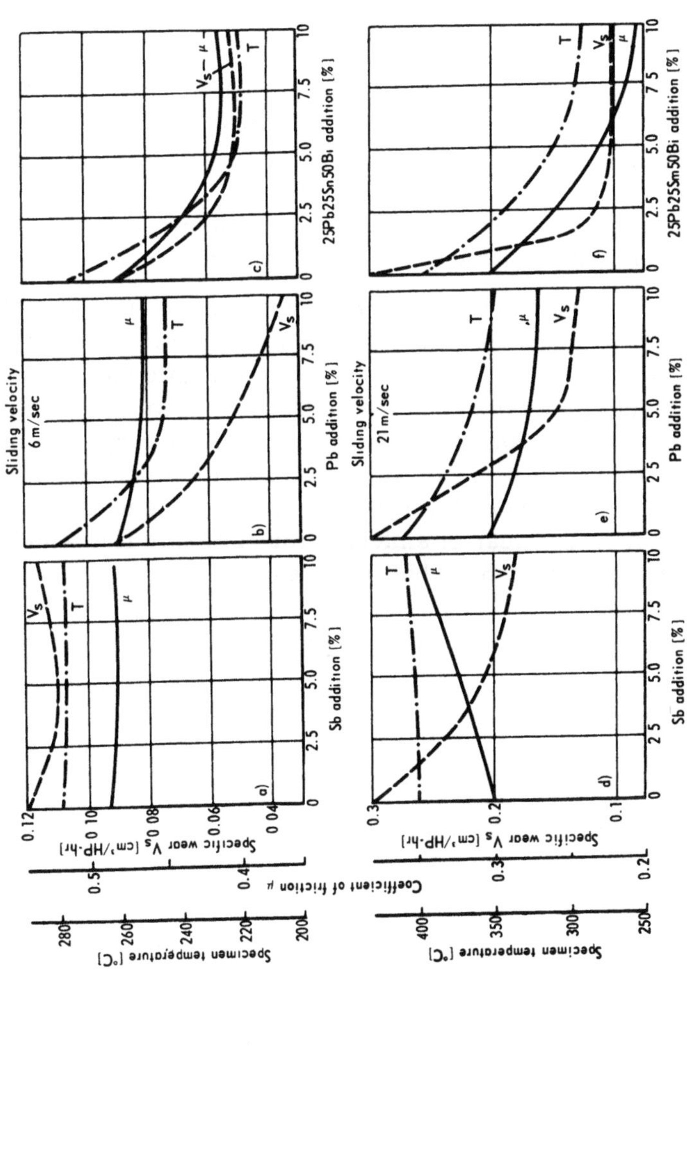

Fig. 7. Mechanical properties and friction characteristics of sintered iron–graphite with the additions of low-melting metals.

in the coefficient of friction. It is interesting that even small additions strongly affect the frictional behavior, whereas further additions exercise only a small influence. The results lead to the conclusion that metallic lubricants can be used, beside graphite, to master the desired frictional properties of sintered iron–graphite.

Conclusion

The investigation of sintered iron–graphite has shown that structures formed by rapid cooling or heat treatment have only a limited influence on frictional behavior. Temperatures developed on the surface during the rubbing process can disrupt or change the structure of iron previously formed, thus voiding whatever effect on the friction properties it might have had.

Further experiments have been carried out to study the effects of metallic additions. The addition of copper to sintered iron–graphite promotes the precipitation of graphite and the formation of ferrite, thus lowering the mechanical properties. Copper additions have significant influence on wear only at very low sliding speeds. Additions of nickel improve strength properties and frictional behavior at very low speeds. Low-melting nonferrous metals can lower the wear of sintered iron–graphite if a liquid phase is formed during the rubbing process.

References

1. Collins, B. T., and Schneider, C. P., International Powder Metallurgy Conference, 1965, New York.
2. Greatorex, J. C., Preprint from the Symposium "Factors Influencing the Economics of the Manufacture of sintered Metal Components, October 21-22, 1965, in London.
3. Kachi Sukeji, International Powder Metallurgy Conference, 1965, New York.
4. Dufek, V., Kovokeramické třecí materiály, SNTL, Prague, 1965.
5. Dufek, V., and Miculek, J., Pokroky prašková metalurgíe, No. 3, pp. 3-11, 1965.
6. Dufek, V.: Preprint from the 3rd International Powder Metallurgy Conference.
7. Procházka, V., Navara, E., and Miškovič, V., Pokroky práškové metalurgie, No. 3, pp. 28-40, 1965.
8. Navara, E., Problemy práškové metalurgie, Bratislava, 1964, pp. 127-140.
9. Strobel, E., Technik, Vol. 12, pp. 819-824, 1957.
10. Strobel, E., Müller, H., and Beer, H., Der Maschinenbau, Vol. 3, pp. 97-101, 1962.

11. Strobel, E., and Uebel, H., Proceedings of the 2nd International Powder Metallurgy Symposium in Eisenach, Akademie Verlag, 1962, pp. 541-550.
12. GDR-Patent DWP 18, 191 and DWP 43, 180 Kl. 40b, 33/02.
13. Strobel, E., Rebsch, H., and Henkel, H., Preprint from the 3rd International Powder Metallurgy Conference.
14. Strobel, E., Rebsch, H., Uebel, H., and Henkel, H., Maschinenbau, Vol. 15, No. 4, pp. 151-156, 1966.
15. Navara, E., Proceedings of the Powder Metallurgy Conference in Cracow, 1963.
16. GDR-Patent DAS No. 1, 183, 697 Kl. 40b-39/4.
17. GDR-Patent DAS No. 1, 208, 082 Kl. 40b-39/54.
18. GDR-Patent DAS No. 1, 186, 658 Kl. 40b-31-04.
19. Bockstiegel, E., Stahl und Eisen, Vol. 79, No. 17, pp. 1187-1201, 1959.
20. Botha, L., Proceedings of the Powder Metallurgy Conference in Cracow, 1963.
21. Domsa, A., Neue Hütte, Vol. 9, No. 3, pp. 186-188, 1964.
22. Benesovsky, F., Berg- und Hüttenmann. Mh., Vol. 96, No. 9, 1951.
23. Findeisen, G., Metall, Vol. 14, No. 10, pp. 995-998, 1960.

Chapter 5

Effect of Graphite Content on the Antifriction Properties of Metallographite Materials

L. I. Pugina, I. M. Fedorchenko, and N. E. Ponomarenko

Academy of Science of the Ukrainian SSR
Kiev, USSR

The development of modern technology, especially turbo-engineering and chemical engineering, requires the creation of materials which will operate without lubrication. Recently, graphite has found increasing use as a solid lubricant for making components intended for operation in dry friction conditions. This material is a constituent part of a series of antifriction carbon–graphite materials impregnated with such metals as bronze, copper, or lead, the purpose of which is to raise strength. The process of making such carbon–graphite materials is complicated and laborious since it requires special equipment.

Powder metallurgy methods enable us to make metallographite materials with a high content of graphite, (for example, electrical brushes on the basis of copper–graphite). These metallographite compositions have low strength. This is due to the fact that the graphite has low mechanical strength, and with increase in its content in the composition, the strength of the material diminishes. Therefore, the content of graphite in this type of composition is normally limited to 5-7%.

Investigations of new antifriction materials on the basis of metal–graphite led to the development of a number of compositions containing 10-30 wt.% of graphite [1], silver–graphite materials for generator brushes in aircraft engines containing graphite in amounts of up to 90 vol.% [2]. For operation without lubricants, recommendations are given for metallographite materials based on brass or bronze for temperatures of from − 200° to + 350°C and materials based on iron for up to + 600°C [3, 4]. Accounts are given [5] on the development of antifriction materials based on graphite containing up to 90 vol.% graphite for bearings working at high temperatures. The metal base of these materials may be Fe, Ni, Co, Cu, Ag, Al, etc. However, the literature contains no data on the use of materials of this type and their behavior with face friction at high speeds.

The aim of this paper was to develop metallographite materials with various graphite contents and to investigate the effect of the latter on the antifriction properties of materials at high speeds.

The feature of the technology for making metallographite materials with high graphite contents consists in sintering in the presence of a liquid phase. Under these conditions the liquid phase is an all-metallic component which provides the best contact with graphite and increases the strength of the material. The graphite is poorly wetted by molten metals and the wettability deteriorates with increase in the content in the metal of dissolved carbon [5, 6]. Introducing special surface-active agents may prevent sweating of the liquid phase during sintering on account of the reduction in the surface energy of the alloy, and give an improvement in the wettability of the graphite. Humenik et al. [5] recommend the following as surface-active agents for iron–graphite: Ce, Ca, Mg, and Ba. The optimum results were obtained by using calcium in the form of a silicon alloy in the ratio of Si : Ca = 65 : 35 in amounts of 0.1%.

The present authors made metallographite materials based on copper–nickel and iron–nickel alloys containing up to 50-80% nickel and having a thermal resistance of up to 673-873°K. The effect of the graphite content on the antifriction properties was investigated within the range 20-90 vol.%. The alloys were made by powder metallurgical methods by a single process of pressing and sintering. The mixtures were prepared without moistening additions (glycerine, alcohol, etc.). Depending on the graphite content, the specimens were pressed at 30-50 kN/cm^2. The porosity of the specimens after pressing was 5-10%. Sintering was done at 1423-1493°K (which developed a liquid phase) with a soak at the optimum sintering temperature of 2 hours.

On the basis of Humenik's data [5] the surface-active additives used were powdered calcium, silica, and a silicon–calcium alloy of 55% Si and 32% Ca (the remainder was iron). The most effective material proved to be the silicon–calcium alloy, but, contrary to Humenick's findings [5], in amounts of not less than 1% of the weight of the metal component of the alloy. Increasing the additions of Si–Ca to 3-5% scarcely affected the properties of the alloy. An addition of less than 1% silicon–calcium alloy proved to be ineffective. After sintering the porosity of the material was 8-13%.

We determined the total macrohardness of the material by the Brinell method with a load of 62.5 kg and sphere of 2.5 mm diameter, carried out a metallographic analysis, and measured the microhardness of the metallic base. For specimens containing 30 vol.% graphite the hardness of the materials with a porosity of 10% lay in the range 40-50 daN/mm^2 and depended very little on the composition of the metallic base. For specimens with about 50 vol.% graphite the hardness of the material dropped sharply and was 18-25 daN/mm^2. With a graphite content of above 70 vol.% (30 wt.%) the material had such a low hardness that it was difficult to measure it with the existing apparatus.

Prepressing of the materials with more than 70% vol. graphite was also difficult, owing to brittleness. Half the specimens broke during prepressing. Prepressing was done on a hydraulic press at a pressure of 30-35 kN/cm^2. After pressing it was not possible to measure the hardness of the specimens. Repeat sintering with the same cycle produced sweating of the metallic phase which is apparently due to the loss after sintering of the surface-active properties of the additives. The brittleness of the material was due to the fact that the metallic phase does not create a continuous framework (Fig. 1). However, specimens of materials containing up to 50-60 vol.% of graphite were amenable to property measurements. Under these conditions, thin-walled rings of 90 mm O. D. and 4 mm wall thickness could be subjected to measurements for ellipse formation and buckling over the height. The materials were also tested using the facial-friction method. The specimens consisted of rings 52 and 75 mm in diameter with a wall thickness of 5-6 mm and a height of 10 mm. Laboratory tests of metallographite materials without lubricants when V = 11.5-42.5 m/sec and P = 0.68 daN/cm^2 showed that specimens containing up to 30 vol.% graphite at friction speeds of 11.5 m/sec heated up quickly and wore out. They were therefore removed from further testing.

Specimens with a high graphite content became heated to a lesser degree, although they had a high coefficient of friction. With increase in the friction rate and also in the graphite content in the material, the coefficient of friction drops slightly, but there is an increase in the temperature of rubbing surfaces and wear of the specimens. This is explained by the crumbling out of the graphite, and the formation of breaks on the working surfaces. All this increases the process of destruction of the material and leads to its dispersion. The test data for specimens with various graphite contents are given in Table 1.

For the tests with lubricants, we used specimens containing 50 vol.% graphite but with different metal bases. The lubricant was fed by the drip method. Under these conditions there was a clear reduction in the coefficient of friction. Table 2 shows that the material can sustain much higher speeds and loads than with dry friction. With maximum friction cycles the material based on copper–nickel–graphite operated for more than 50 hr (when V was

Table 1. Results of Tests on Metallographite Specimens with Different Graphite Contents When V = 11.5 m/sec and P = 0.68 daN/cm^2 with Dry Friction (Standard material was steel U9A, HRC = 52)

Standard material index	Graphite, vol.%	Wear of specimen hr/mm	Coefficient of friction	Temperature close to friction surface	Remarks
1	50	0.280	0.36	60	Series of specimens
2	70	0.90	0.40	45	based on copper
3	80	0.107	0.22	40	–nickel–graphite
4	90	0.015	0.27	55	
5	50	0.036	0.39	60	Series of specimens
6	70	0.72	0.31	45	based on iron
7	80	0.006	0.26	95	–nickel–graphite
8	90	0.030	0.35	90	

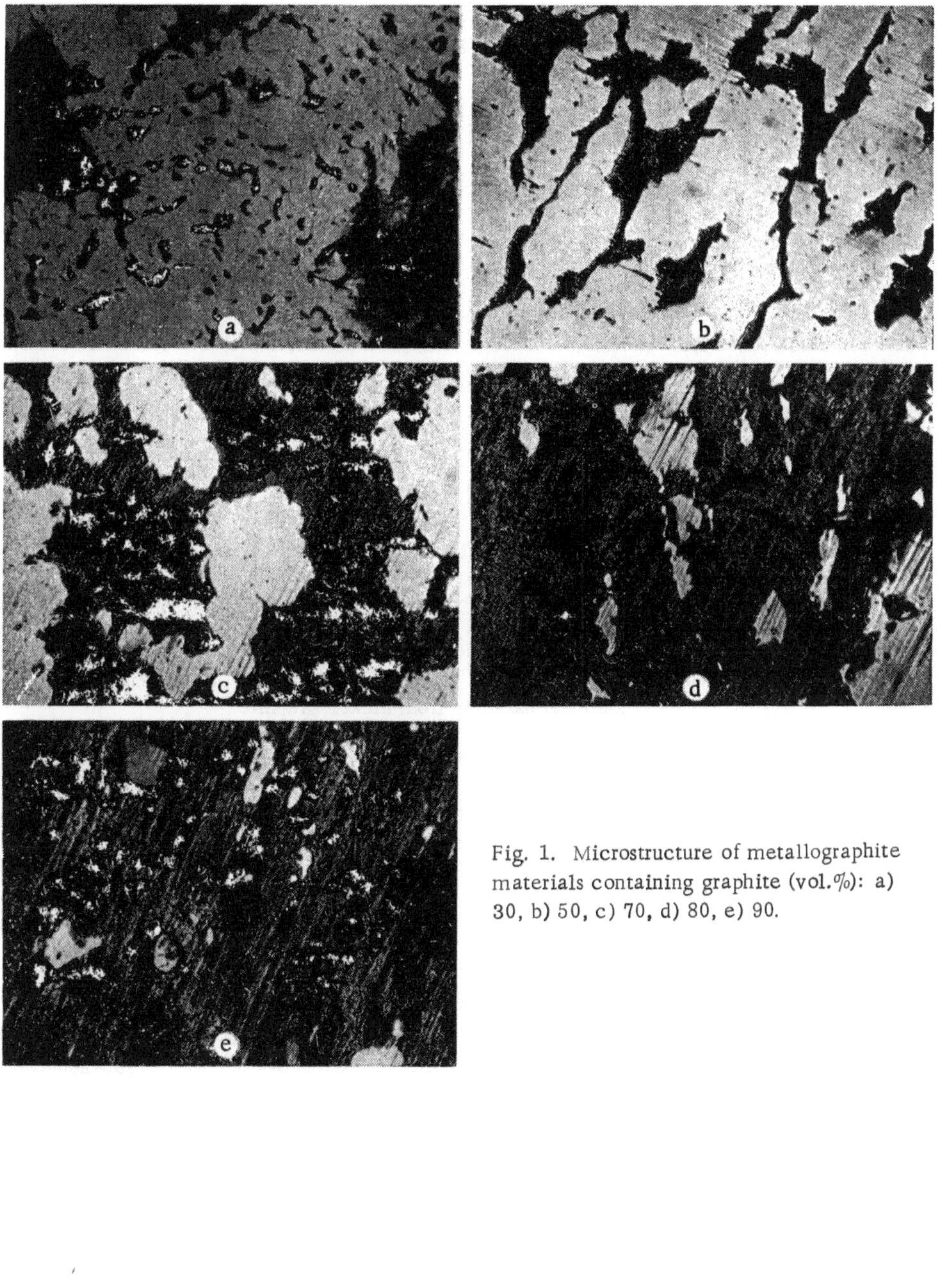

Fig. 1. Microstructure of metallographite materials containing graphite (vol.%): a) 30, b) 50, c) 70, d) 80, e) 90.

Table 2. Data on the Test Results of Certain Metallographite Materials with Lubrication during Facial Friction (Standard material was steel U9A, HRC = 52)

Friction speed, m/sec	Specific pressure, daN/cm²	Wear time, min	Wear per hour, μ	Coefficient of friction	Temperature close to friction surface, °C
Material based on copper–nickel–graphite					
11.5	0.55	60	none	0.105	65
32.0	0.55	10	"	0.09	64
49.5	0.55	10	"	0.065-0.05	80
11.5	0.82	60	0.003	0.07	65
32.0	0.82	10	0.002	0.06	70
49.5	0.82	10	none	0.05	84
11.5	1.65	60	"	0.052	79
32.0	1.65	10	"	0.057-0.062	100
49.5	1.65	10	"	0.052	110
11.5	3.03	60	"	0.050	100
32.0	3.03	10	"	0.050	135
49.5	3.03	10	"	0.04	142
21.5	3.3	60	"	0.04	142
32.0	3.3	10	"	0.04	150
49.5	3.3	10	"	0.04	170
21.5	3.86	60	0.001	0.045	140
32.0	3.86	10	none	0.045	154
49.5	3.86	10	"	0.04	250
Material based on iron–nickel–graphite					
11.5	0.27	240	none	0.18-0.26	55
11.5	0.82	180	0.0003	0.1-0.12	88
32.0	0.82	10	none	0.05	72
32.0	1.65	10	0.006	0.052	104
49.5	1.65	10	0.012	0.052	115
32.0	2.2	10	none	0.056-0.064	144
49.5	2.2	10	"	0.046-0.05	180
49.5	2.48	10	"	0.048-0.054	190
49.5	2.75	10	"	0.3	118
49.5	3.03	10	"	0.028-0.032	115
49.5	3.3	10	"	0.026-0.03	116
49.5	3.56	10	"	0.044-0.078	250-280

as high as 49.5 m/sec, P up to 3.8 daN/cm^2, PV up to 195 $kgm/cm^2 \cdot$ sec), and materials based on iron–nickel–graphite for more than 40 hr (with V up to 49.5 m/sec, and P up to 3.3 kg/cm^2, PV up to 165 $kgm/cm^2 \cdot sec$). Under these conditions the wear of the specimens was almost unnoticeable, and the coefficient of friction lay in the range 0.03-0.06, depending on the velocity, load, and composition of the material. Further increase in the load and velocities above the indicated limits led to overheating of the specimens and burning out of the material.

Summary

Metallographite materials containing not more than 50 vol.% graphite can be used for rubbing-friction applications at high velocities in conditions of limited lubrication. Further increase in the graphite content sharply reduces the strength of the material.

With increase in the friction rate and loading, the coefficient of friction diminishes. This tendency occurs also with increase in the graphite content, when the reduced strength of the material leads to an increase in wear.

The metallographite materials investigated (containing 50-90 vol.% graphite) under dry friction conditions have high coefficients of friction (0.22-0.40), which means that they cannot be recommended as antifriction materials.

For operation with a lubricant the metallographite materials can operate as antifriction materials with coefficients of friction of 0.03-0.06 at velocities up to 50 m/sec and with loads up to 3.3-3.8 $kgm/cm^2 \cdot sec$ (PV = 165-195 $kgm/cm^2 \cdot sec$).

References

1. Aizenkol'b, F. et al., East German Patent No. 18191 (February 2, 1960).
2. US patent No. 2537591, Application, October 10, 1945; January 9, 1956.
3. Compressed Air and Hydraulics, Vol. 25, No. 294, p. 365, 1960.
4. Terent'ev, Ya. K., Énergomashinostroenie, No. 12, p. 45, 1961.
5. Humenik, M., Hall, D. V., and Alsten, R. L., Metall Progress, Vol. 81, No. 4, p. 101, 1962.
6. Naidich, Yu. V., and Kolesnichenko, G. A., Poroshkovaya Metallurgiya, No. 6, 1961.

Chapter 6

The Influence of Silica and Alumina Additions to Sintered Iron-Based Friction Materials

Václav Procházka and Erik Navara

Laboratory of Metallurgical Technology
Technical University
Košice, Czechoslovakia

Introduction

Sintered iron-based friction materials are expected to perform reliably up to the temperature of 1000°C or even more without substantial or sudden deterioration of frictional properties. One of the main requirements is, therefore, thermal stability, which means that the values of friction coefficient and wear rate do not appreciably change up to the specified maximum temperature [1, 2]. The present state of the theory of friction of metals does not quite exactly explain conditions which ensure thermal stability or, on the other hand, causes for thermal instability of many friction materials. Contemporary theories and known laws governing metallic friction, worked out mainly by Bowden [3], Goodzeit [4], Coffin [5] and Semenov [6], make it possible to select materials with very little wear and suitable coefficient of friction when cold, yet they fail to explain changes in the behavior of friction materials at high temperatures.

Kosteckij [7] proposes a new distinction of types of wear, which takes into account high-temperature wear. He distinguishes five types of wear: oxidation, fatigue, abrasive wear, and surface damage at low and high temperatures. Surface damage at low temperatures is identical with the disruption of metallic junctions formed by cold or pressure welding, which is a well-known and thoroughly studied phenomenon. Surface damage at high temperatures, or temperature wear according to Kosteckij, depends upon the state of thermal plasticity of the surface. This expression lacks an accurate definition, because experimental work does not yet enable us to draw decisive conclusions.

A number of studies and an appreciable amount of research has been carried out regarding cold welding and wear of sintered steel [8, 9]. The results, however, do not give sufficient information for the development of sintered material designed to withstand high temperatures. Nevertheless, plenty of evidence has been collected by experience that several nonmetallic additives, mainly barium sulfate and also so-called abrasive additives such as silica or alumina, improve friction properties at high temperatures and promote heavy-duty performance. The effect of these additives is by no means simple. Barium sulfate, for example, decomposes during sintering. As regards silica and alumina, no attention has been paid to their chemical stability in the presence of carbon. As there are significant differences in their effects, we have made an attempt to discover the causes.

Previous Work

Our earlier experiments have led to the conclusion that graphite and barium sulfate are suitable additives to heavy-duty, iron-based friction material. This is acknowledged by a number of authors; however, their opinions as to the respective amounts of these nonmetals vary considerably. We have found that high graphite contents (16-24%) are more suitable for low-temperature performance; but in conditions of poor heat transfer, the addition of graphite should not exceed 6 to 8%. The resulting deterioration of frictional properties is efficiently compensated by barium sulfate. It has been proved at this laboratory that barium sulfate undergoes complete reduction by carbon during sintering according to the equation

$$BaSO_4 + 4C = BaS + 4CO$$

The effect of addition of barium sulfate was tested up to the amount of 18 wt.%. It was found that resistance to thermal wear increased with barium sulfate content. On the other hand, larger additions of barium sulfate led to a rapid decline of mechanical properties. For instance, compressive strength decreased from 68.8 kp/mm^2 at 12% $BaSO_4$ addition to 28 kp/mm^2 at 18%. Brinell hardness decreased from 44 to 30.5 kp/mm^2 at the same time. As a result of an analysis of a series of mechanical as well as frictional measurements, the addition of $BaSO_4$ was fixed to 14%, together with graphite addition of 6%. Other investigated nonmetals were always added to this standard mixture.

Experimental Procedure

Specimens prepared for frictional tests were formed into a cylinder of 30 mm diameter and approximately 10 mm height. They were pressed at 5 Mp/cm^2, the amount of powder being kept constant. All mixtures were prepared from the following raw materials:

1. Iron powder produced by eddy mills, annealed in hydrogen at 650°C, particle size – 200 mesh;
2. Natural graphite, dried at 150°C, particle size – 200 mesh; ash content 12%; ash composition: 47.15% SiO_2, 41% Fe_2O_3, 5.42% CaO, 1.9% MgO, balance unidentified;
3. Natural barium sulfate purified by flotation, particle size – 200 mesh, calcinated in air at 700°C. Barium sulfate content 95.4%, impurities mainly SiO_2 (1.2%), Fe_2O_3 (1.6%);
4. Silica, particle size – 200 mesh, dried at 180°C,
5. Alumina, particle size – 200 mesh, calcinated at 1300°C.

Individual mixtures were prepared by mixing in a rotary drum 4 hours. Pressed compacts were sintered in a retort furnace by the following technique. During temperature rise up to 700°C, dried hydrogen was passed through the retort. When the said temperature was attained, hydrogen was replaced by nitrogen. The sintering temperature was 1000°C, sintering time 2 hours.

Sintered compacts were bored to obtain annular specimens. These were then tested on the friction apparatus SFM-2, which had been built at our institute for the purpose of testing heavy-duty materials upon the same principle as the widely used Soviet apparatus I-47[1]. This apparatus is designed for the investigation of friction materials at high temperatures. The area and form of

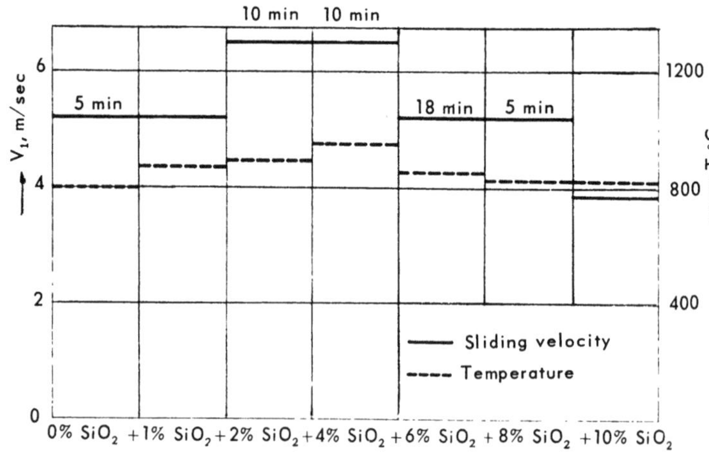

Fig. 1. Frictional thermal stability as a function of the amount of silica added.

sliding surface of the specimen, as well as of the counterpiece, are the same, as a result of which the heat transfer from the sliding surfaces is very poor. Thus high temperature is generated in the friction area at relatively low load and sliding velocities.

Before the actual test, the sliding surfaces were run in at low speed until the initial periodical fluctuation of the friction coefficient ceased. Then the specimen was tested in the whole range of revolutions at the following mean peripheral velocities: 0.13, 0.26, 0.63, 0.91, 1.3, 1.88, 2.6, 3.25, 3.9, 5.2, and 6.5 m/sec. The test load was constant and equaled 15 kp/mm^2. The adjusted velocity was maintained for 15 minutes, and measurements of friction coefficient and temperature were taken every 5 minutes. The friction temperature was measured by a thermocouple, the end of which was inserted into a hole 1.2 mm in diameter bored into the counterpiece. After the complete test, the bulk and linear wear were found by weighing and precise measuring. In case the specimen did not withstand the whole test due to excessive wear, the time of run at the highest speed was recorded.

Effect of Silica

Addition of silica to the standard mixture containing 6% graphite and 14% barium sulfate, balance iron, was tested in the range from 1 to 10%. Pressing and sintering were carried out as described above.

Tests on the SFM-2 apparatus have shown that thermal stability increases up to the amount of 4% SiO_2; further rise of silica content leads to a decrease of thermal wear resistance. As shown by Fig. 1, specimens containing 2 and 4% SiO_2 endured 10 min at the highest speed of 6.5 m/sec; the temperature of the first specimen rose to 890°C, the second attained 950°C. Higher SiO_2 content – 6% – leads to a marked thermal wear already at a lower testing speed of 5.2 m/sec, which the specimen withstands for only 10 min. The test time at the said velocity is decreased to 5 min with 8% SiO_2, and specimens with 10% silica fail already at 3.8 m/sec, when the temperature reaches 830°C.

For the sake of comparing the frictional properties of specimens with different nonmetallic additives, two sliding velocities were selected: 0.26 and 2.6 m/sec. The first velocity with regard to low-temperature wear occurs markedly at low sliding speeds, the second is expected to give rise to thermal wear. Increasing content of silica raises the coefficient of friction and temperature at both velocities, as indicated in Fig. 2. At 0.26 m/sec the coefficient of friction is 0.46 when silica content equals 1%, and is raised to 0.66 when silica content amounts to 10%; the temperature changes at the same time from 125°C to 195°C. Rather similar is the influence of the silica at a speed of 2.6 m/sec: the coefficient of friction is 0.2 at 1% SiO_2 and 0.28 at 10% SiO_2, with the temperature reaching 410°C in the first case and 606°C in the latter.

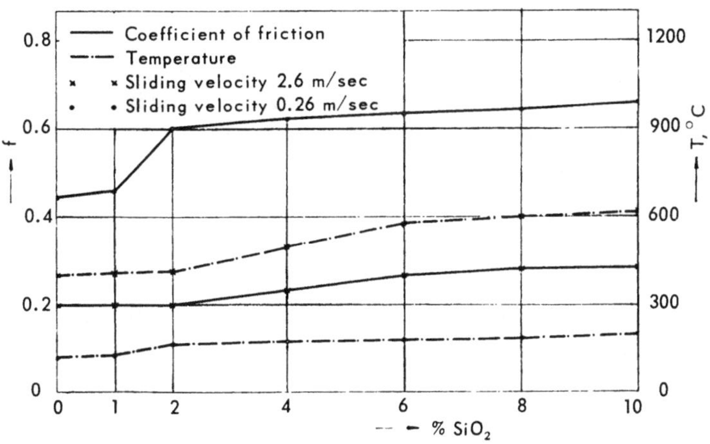

Fig. 2. Effect of silica upon the frictional properties at the selected velocities of 0.26 and 2.6 m/sec.

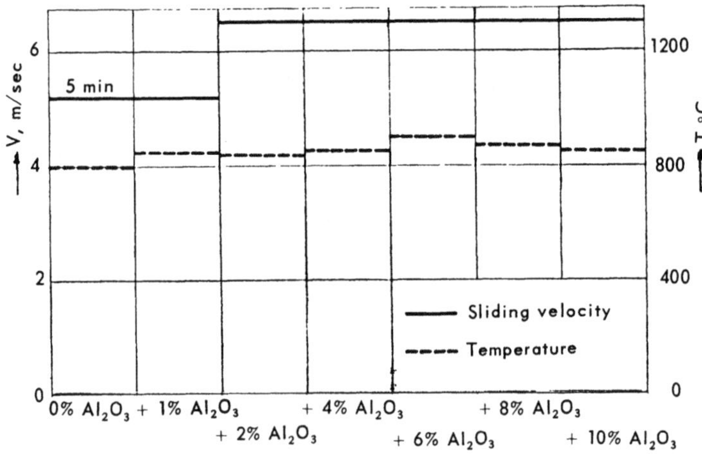

Fig. 3. Frictional thermal stability as a function of the amount of alumina added.

The silica content also affects mechanical properties, such as compressive strength and Brinell hardness. Compressive strength falls from 58.5 kp/mm^2 when there is only 1% SiO_2 to 25 kp/mm^2 when SiO_2 content amounts to 10%. The decrease of hardness is from 40.5 kp/mm^2 to 35.5 kp/mm^2.

Effect of Alumina

The additions of alumina were the same as silica, i.e., 1%, 2%, 4%, 6%, 8%, and 10% to the standard mixture; pressing and sintering were performed as described.

Friction tests have shown that thermal stability increases rapidly with small alumina additions and no negative influence of larger amounts of alumina was found. As Fig. 3 indicates, 1% Al_2O_3 prolonged the test time at a velocity of 5.2 m/sec up to a limit of 15 minutes; yet the specimen failed quickly at the highest speed. However, specimens with 2% of alumina or more passed the complete test, at which the temperature varied between 840°C to 900°C.

The friction coefficient at low speed – 0.26 m/sec – is almost unaffected by alumina, and lies within a narrow range 0.49 to 0.52 (see Fig. 4). The temperature rises moderately. At the higher velocity, 2.6 m/sec, the influence of Al_2O_3 content is also negligible. The friction coefficient increases from 0.17 at 1% to 0.25%

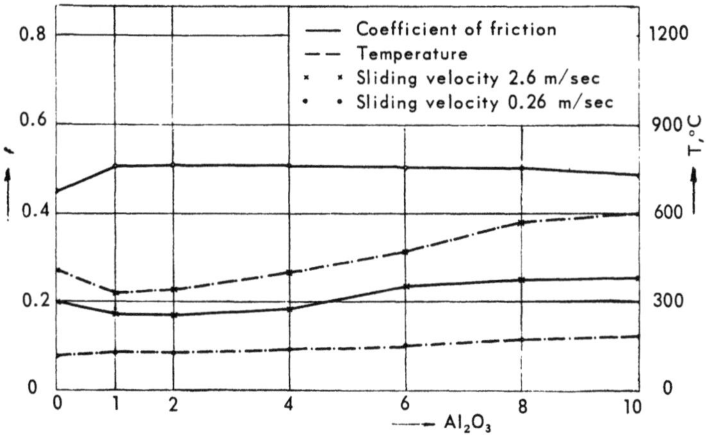

Fig. 4. Effect of alumina upon the frictional properties at the selected velocities of 0.26 and 2.6 m/sec.

at 10% Al_2O_3. The temperature rises at the same time from 330°C to 595°C.

Compressive strength decreases as with other nonmetallic additives. The value of 48 kp/mm² for specimens with 1% Al_2O_3 falls to 21.4 kp/mm² when the alumina content equals 10%. Hardness changes very slightly. Up to 4% alumina, there is little decrease (from 41 to 38 kp/mm²); further increase of alumina brings about a small increase of hardness up to 41.3 kp/mm² at 10% Al_2O_3.

Changes of Nonmetallic Additives during Sintering

Differences in the effect of silica and alumina upon mechanical and frictional properties lead to the assumption that chemical changes may occur during sintering. Chemical analyses of silica-containing specimens have proved that silica is partly reduced into silicon, which dissolves in iron. Results of the analyses are given in Table 1. The table illustrates not only partial reduction of silica but also a decrease in the carbidic carbon content. This is supposed to be caused directly by the dissolved silicon, the graphitization effect of which is well known.

The photomicrographs in Figs. 5 and 6 illustrate the decrease of cementite content as a result of higher addition of silica. The first micrograph shows the structure of a specimen with 4% silica,

Table 1. Chemical Analyses of Sintered Friction Materials with Different Silica Contents

Addition of silica, %	Silica from added graphite, %	Free SiO_2, %	Dissolved Si, %	Relative amount of reduced Si, %	Carbon, total, %	Carbon, chemically bonded, %
0	0.4	0.02	0.17	92.5	2.77	1.43
1	0.4	0.25	0.53	82.0	2.74	1.42
2	0.4	1.0	0.65	58.3	2.7	1.40
4	0.4	1.6	1.3	63.6	2.6	1.36
6	0.4	2.7	1.7	57.8	2.65	1.30
8	0.4	3.9	2.1	53.5	2.7	1.22
10	0.4	5.8	2.15	44.3	2.7	1.0

where large grains of cementite are to be seen; the second, prepared from a specimen with 10% silica addition, displays only a thin and discontinuous network of cementite at grain boundaries.

The reduction of silica may be attributed to carbon monoxide, according to the equation

$$SiO_2 + 2CO = Si \text{ (dissolved in Fe)} + 2CO_2$$

Silicon dissolves in iron and carbon dioxide reacts immediately with free carbon, thus regenerating the monoxide:

$$CO_2 + C = 2CO$$

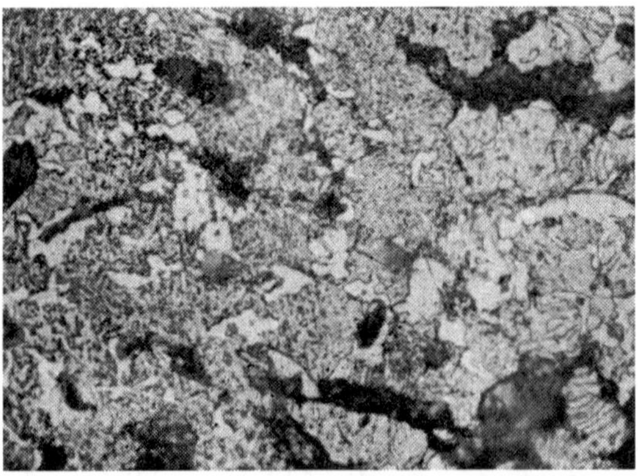

Fig. 5. Photomicrograph of specimen containing 4% SiO_2 (500×).

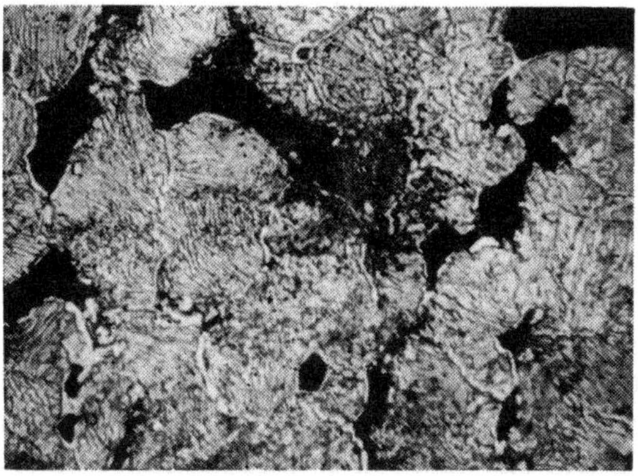

Fig. 6. Photomicrograph of specimen containing 10% SiO_2 (500×).

This mechanism explains the possibility of silica reduction at a relatively low temperature, about 1000°C. The rate of the reduction is slow, yet sufficient to account for the measured data.

Even direct reduction by carbon can play its part, although, according to thermodynamic data, the pure system SiO_2 + C requires 1590°C. However, the presence of iron diminishes the activity of silicon and, as such, supports the reduction. The schematic reaction

$$Fe + SiO_2 + 2C = FeSi + 2CO$$

may therefore occur at a considerably lower temperature.

Table 2. Chemical Analyses of Carbidic Carbon in the Specimens of Friction Materials with Different Alumina Contents

Addition of alumina, %	Carbon, chemically bonded, %
1	1.43
2	1.34
4	1.37
6	1.33
8	1.37
10	1.38

In contradistinction to silica, alumina remains stable during sintering. Direct reduction by carbon is possible only at 2000°C and the rate of indirect reduction by carbon monoxide is far slower than that of silica, and very likely negligible at 1000°C. Thermodynamic data, however, do not make it possible to decide whether Al_2O_3 remains untouched in the presence of nitrogen. The free enthalpy of the synthesis of Al_2O_3 is about −200 kcal at 1000°C but the affinity of aluminum to nitrogen is also very large, as indicated by the free enthalpy of nitride formation, which is −82 kcal at 1000°C. The presence of aluminum nitride in sintered friction materials has not been proved, but they may contain at least some traces of it. The addition of alumina has no effect upon the carbidic carbon content, as revealed by chemical analyses in Table 2.

This fact is demonstrated by micrographs in Figs. 7 and 8 illustrating the structure of specimens with 4% and 10% alumina additions. It is evident that the cementite content is practically the same in both cases.

Discussion of Results

Both the investigated oxides improve frictional properties, especially thermal stability. However, there are significant differences between their respective effects. Silicon oxide lowers the thermal conductivity of the material, because of its reduction and dissolution of silicon in iron. That is one of the reasons for

Fig. 7. Photomicrograph of specimen containing 4% Al_2O_3 (500×).

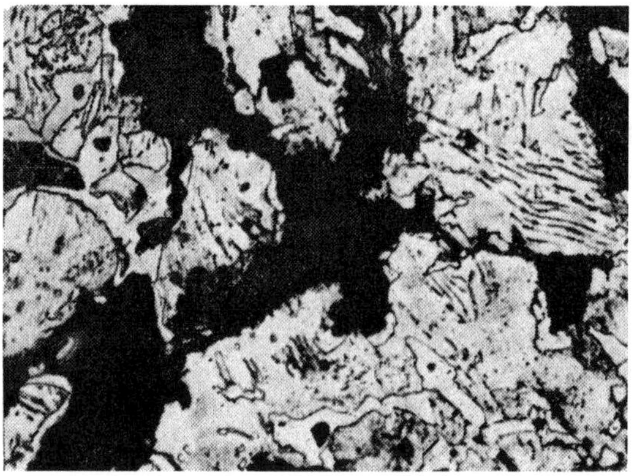

Fig. 8. Photomicrograph of specimen containing $10\%\ Al_2O_3$ (500×).

higher surface temperatures, compared to those of alumina-containing specimens. Figure 1 shows the increase of temperature plotted against silica content. The specimen with 10% SiO_2 reaches the temperature of 800°C at a relatively low sliding velocity of 3.9 m/sec.

Another negative influence of silica lies in the fact that dissolved silicon lowers the percentage of carbidic carbon, which deteriorates the frictional properties especially at high temperatures.

Without further investigation of the respective effects of the two causes, it has been empirically found out that deterioration of frictional properties begins when silica addition exceeds 4%. Aluminum oxide, on the other hand, has no measurable influence upon the metallic matrix and, at the same time, increases the thermal stability. Specimens with 2%-10% Al_2O_3 withstood the complete test in the SFM-2 apparatus, as shown in Fig. 3. Due to its chemical stability which is superior to that of silica, alumina appears to be a more suitable additive to iron-based friction materials.

References

1. Kragelskij, V. V., Friction and Wear, Mashgiz, Moscow, 1962 (in Russian).
2. Dufek, V., Sintered-Metal Friction Materials, SNTL, Prague, 1965 (in Czech).
3. Bowden, F. P., and Tabor, D., The Friction and Lubrication of Solids, Oxford, Clarendon Press, 1954.

4. Goodzeit, C. L., "Selecting bearing materials that will not seize," Mats. Des. Engng., Vol. 47, pp. 105-109, 1958.
5. Coffin, L. F., "Some metallurgical aspects of friction and wear," in: Friction and Wear, Elsevier., Amsterdam, 1959, p. 36.
6. Semenov, A. P., "The phenomenon of seizure and its investigation," Wear, Vol. 4, pp. 1-9, 1961.
7. Kosteckij, V. I., Wear Resistance of Machine Elements, Mashgiz, Moscow, 1961, (in Russian).
8. Navara, E., "The investigation of frictional properties of various sintered-steel structures," in: Problems in Powder Metallurgy, VSAV, Bratislava, 1964, pp. 127-140 (in Czech).
9. Procházka, V., and Miškovič, V., "The contribution to the investigation of dry friction of sintered steel," Kovové materialy, pp. 408-418, 1964 (in Czech).

Chapter 7

Experience in Manufacturing Iron-Base Sintered Friction Alloys

L. P. Lebedeva, B. T. Krysin, Ya. V. Kolpakov, L. N. Ignatov, V. A. Mikhailovskii, G. G. Smirnov, and M. V. Tsytsenko

For the charge used in making iron-base sintered friction parts the following powders are employed: PZh-1, PZh-2, and PZh-3 iron powders (All-Union State Standard 9849-61) and PM-1, PM-2 copper powders (All-Union State Standard 4960-49). In the as-delivered state all these brands of powder contain a certain amount of bound oxygen in the form of oxide films and impurities. Under the All-Union State Standards mentioned above, the oxygen content in iron powders brand PZh-1 and PZh-2 should not exceed 0.5 and 1.0%, respectively, but no regulations are laid down with regard to PZh-3 or to the copper powders. As shown by mass analysis, the oxygen content in PZh-3 powder ranges from 1.1 to 1.7%, and in the copper powder from 0.1 to 0.2%.

The effect of the degree of oxidation of the powders on the manufacturing technology and physicomechanical properties of the parts has been described in a number of papers [1-3]. The investigations can be reduced basically to the following:

1. The oxidation of iron powder containing up to 1% oxygen has no effect on pressibility. When the oxygen content is raised above 1%, pressibility is impaired through a decline in the ductility of the powders.

2. Given identical specific pressing pressures, the porosity of compacts made from the oxidized powders is greater than that of those made from prereduced powders. According to Bebnev [2], an increase of more than 60% in the specific pressure when working with oxidized powders makes it impossible to obtain pressings with a density corresponding to that of pressings made from prereduced powders. The porosity of the pressings made from oxidized powders is approximately 1.5 times greater. Below we give data on porosity as a function of the pressing pressure and oxidation of metallic iron:

Content* of metallic iron, %	94	86	72
Pressing pressure, daN/mm^2	44	55	71
Porosity, %	22	30	30

*Content of metallic iron shows degree of oxidation of powder.

3. When compacts made of oxidized iron powders are sintered, their shrinkage is greater than in compacts made of prereduced powders.

4. The sintering of oxidized compacts, i.e., combination of the sintering operation with reduction, improves the mechanical properties of the parts, e.g., the tensile strength, elongation, and impact strength.

Thus, the presence of oxides in the powder impairs their pressibility and at the same time intensifies the sintering process. This holds for all oxides that are easily reduced by hydrogen (iron, copper, nickel, and other oxides).

To improve the technological properties of the powders, first and foremost their pressibility, the manufacturing process for sintered friction parts includes prereduction of iron as well as copper powders. This operation is carried out in muffle or shaft-type electric furnaces in a hydrogen medium at 400-500°C in the case of copper powder, and at 600-700°C for iron powder. The prereduction time is 3 hr. The oxygen content of the prereduced iron powder is not more than 0.5%, and for the copper not more than 0.05%.

Prereduction of the powders is a labor-consuming process that requires special equipment and instruments, plus additional operations such as grinding up the sintered powders after prereduction and screening into fractions of the required grain size. Furthermore, these additional operations in preparing the charge materials create a dust-laden atmosphere in the workshop, giving rise to unhygienic conditions for the workers and an additional loss of material (screening-out).

Since friction parts are simple in geometric design, i.e., plates 3.0-7.0 mm thick, it was assumed that the use of nonprereduced powders would not cause difficulty during pressing, but would improve the sintering process, and that the prereduction operation might be discontinued. To check this assumption, during our research we investigated the effect of the degree of oxidation of the powders on the technological and frictional properties of iron-base sintered parts.

The oxygen content of iron and copper powders was determined from the loss in weight of a powder sample reduced with dried hydrogen. A portion of powder weighing 5-10 g was placed in a porcelain or quartz boat inside an electric tube-type furnace and calcined in a dry hydrogen medium at 900-950°C in the case of iron, and at 580-600°C in the case of copper. The calcining time was 50-60 min.

The degree of oxidation of the copper powder in the as-delivered state is only slight (up to 0.20% oxygen) and has no effect on its technological properties. It is also extremely small compared with the degree of oxidation of iron powder, and for this reason cannot have any appreciable effect on the overall oxygen content in the charge.

We made up 5 batches of charge, the composition of which included iron powders with the following degree of oxidation:

Batch No.	O_2 content, %
1	0.7
2	1.5
3	2.0
4	3.5
5	5.4

The charge was composed of the following powders: 1) iron powder type PZhZM. All-Union State Standard 9849-61; 2) copper powder mark PM-2, All-Union State Standard 4960-49; 3) graphite mark KLZ, All-Union State Standard 5279-61; 4) barytes concentrate (commercial), All-Union State Standard 4682-49; 5) serpentine asbestos (chemically pure sheets), All-Union State Standard 2850-58; 6) Lyuberets molding sand, All-Union State Standard 2138-56.

The graphite and sand were added to the charge after drying for 4 hours at 120-150°C and screening through a 01 mesh for graphite and a 016 mesh for the sand. The barytes concentrate and asbestos were calcined for 4 hours at 700°C in the first case, and for 3 hours at 950-1000°C in the second; they were ground down in a ball mill and screened through a 0125 mesh. The copper powder was added to the charge in the as-delivered state, containing 0.20% oxygen.

Different amounts of oxygen in the iron powders were obtained by oxidizing PZhZM powder, which in the as-delivered state contains 1.5% oxygen. To oxidize the powders they were heated in electric furnaces, without a protective medium, under the conditions described in Table 1.

To produce iron powder containing 0.7% oxygen, PZhZM powder was prereduced under the conditions described above. The given charge components were mixed for 4 hours in a "drunken barrel" mixer. Specimens were pressed from the charge in the form of plates 5.0 mm thick with an area of 34 cm^2.

A charge portion weighing 81 g and specific pressing pressure 60 kN/cm^2 were adopted for all the variations studied, enabling us to obtain a density of 5.2-5.3 g/cm^3 for specimens of the charge containing prereduced iron powder. The charge was pro-

Table 1

Oxygen content in powder, %	Heating temperature, °C	Soaking time, hr
2.00	250	4
3.50	300	4
5.40	350	4

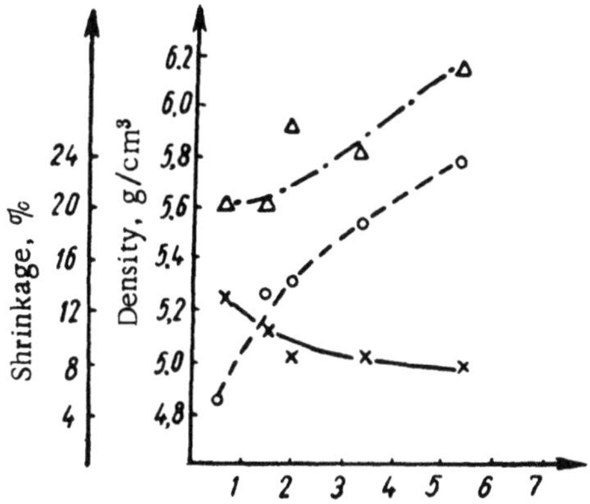

Fig. 1. Variation in density and shrinkage as a function of the oxygen content in iron powder: x—x density of pressing; △—·—△ wear and tear of ChNMK pig iron; ○---○ shrinkage.

portioned by the weight method. A unilateral pressing system was adopted for the specimens made in a hydraulic press.

The specimens were sintered under pressure in a shaft-type electric furnace in a hydrogen medium at 1060 ± 10° and soaked for 2.5 hours. The specific pressure during sintering was 16 daN/cm^2. Twenty-four specimens were made from each charge containing different amounts of oxygen. The pressibility of the charge, density and shrinkage of the specimen during sintering, and its hardness, chemical composition, and frictional properties were then determined.

Effect of Degree of Oxidation of Iron Powder on Pressibility of Charge. As mentioned above, all the batches of specimens were pressed at the same pressure. The pressibility was evaluated from the density of the specimens obtained: the higher the density, the better the pressibility. The density of the compacted specimens was determined theoretically. Figure 1 shows the density of the specimens (mean of four determinations) for different oxygen contents in the iron powder.

As can be seen from the data given, as the degree of oxidation of the iron powder is increased, the density of the pressings falls steadily. When the oxygen content is raised from 0.7 to 2.0%,

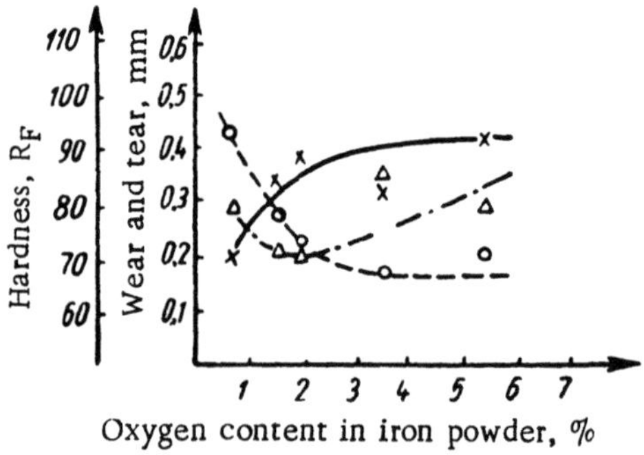

Fig. 2. Wear and tear and hardness as functions of the oxygen content in iron powder: ×—× hardness; △−·−△ wear and tear (of ChNMK pig iron); ○---○ = wear of sintered alloy.

the density declines from 5.2 to 5.02 g/cm^3. A further increase in the degree of oxidation leads to a slight decrease in density. When the oxygen content in the powder is 5.4%, it amounts to 4.93 g/cm^3. The moldability of the specimens is satisfactory. After pressing, the specimens showed no striation or crumbling of the faces.

Effect of Degree of Oxidation of Iron Powder on Density and Shrinkage during Sintering. The linear shrinkage (%) was calculated from measurements of the height of the specimens before and after sintering. Shrinkage data for the specimens (Fig. 1; mean of seven determinations) show that when the oxygen content is increased in the powders to 5.4%, the shrinkage increases from 6 to 23%, i.e., almost by a factor of 4. As the oxygen content in the iron powder is increased, the compactability of the charge deteriorates, while the density of sintered specimens increases through greater sintering shrinkage. For an oxygen content of 0.7% in the charge, the density is 5.6 g/cm^3, and for 5.4% oxygen it is 6.1 g/cm^3.

Effect of Degree of Oxidation of Iron Powder on Hardness. Hardness was measured at three places on each specimen along the diagonal. It was determined in 25 specimens using a Rockwell tester with a $^1/_{16}$-inch ball and a load of 60 kg (scale F). The hardness value was taken as the arithmetic mean of three determinations.

Figure 2 gives data for the variation in hardness as a function of the oxygen content in the iron powder. Each of the points on the graph corresponds to the mean hardness, ascertained with 25 specimens. It follows from this diagram that as the oxygen content in the iron powder is raised from 0.7 to 2.0%, the hardness of the specimens increases by 19 units (27%), but when the oxygen content is further increased, it hardly varies at all.

Effect of Degree of Oxidation of Iron Powder on Frictional Properties. Using an I-47 tester and the techniques developed by the Institute of Machine Study for a shorter test-period, we determined the frictional properties of the specimens. The tests were carried out with hollow cylinders with an outside diameter of 28 mm and an inside diameter of 20 mm at a specific pressure of 15 daN/cm^2. The sintered alloy specimens were tested in pairs with specimens of ChNMKh pig iron possessing a Brinell hardness of 160-200 daN/mm^2.

The frictional thermal stability of the material was assessed from the friction coefficient at different temperatures occurring during the experiment, and from the linear wear and tear of the specimens during the testing cycle. The degree of wear on the sintered and pig iron specimens containing different amounts of oxygen in the iron powder are shown in Fig. 2. Each point on the graph represents the mean value of the wear and tear for four specimens. When the oxygen content in the iron powder is varied 0.7 to 5.4%, the wear of the sintered alloy is approximately halved. As the oxygen content is raised to 2%, the wear of the alloy is only 0.23 mm, i.e., 47% less than the wear in specimens containing 0.7% oxygen (0.44 mm).

Wear of the pig iron specimen when the oxygen content is increased from 0.7 to 1.5-2.0% is reduced to a minimum, and when

Table 2

Oxygen content in iron powder, %	Friction coefficient at temperature, °C									
	100	200	300	400	500	600	700	800	900	1000
0.7	0.63	0.57	0.49	0.43	0.37	0.32	0.26	0.23	0.20	0.20
1.5	0.63	0.56	0.48	0.45	0.39	0.36	0.29	0.26	0.22	0.22
2.0	0.63	0.56	0.49	0.45	0.43	0.39	0.29	0.25	0.21	0.19
3.5	0.63	0.58	0.52	0.45	0.41	0.34	0.28	0.25	0.23	0.21
5.4	0.61	0.56	0.49	0.46	0.44	0.39	0.32	0.26	0.24	0.20

Table 3

Oxygen content in iron powder, %	Carbon content in specimens, %	
	prior to sintering	after sintering
0.7	6.38	5.69
1.5	6.91	5.62
2.0	6.37	4.50
3.5	5.54	3.74
5.4	5.85	3.04

it is further increased (up to 5.4%), the wear approximately equals that of specimens containing 0.7% oxygen. Table 2 gives the friction coefficients obtained (mean of four determinations). As can be seen from the table, the friction coefficient at low temperatures (100, 200, and 300°C) is virtually independent of the oxygen content in the charge, but at 400-900°C we note a certain increase in the friction coefficient for increasing oxygen content in the charge.

Effect of the Degree of Oxidation of Iron Powder on the Chemical Composition of the Sintered Mixture. Table 3 gives the carbon content in specimens made from iron powders with different degrees of oxidation, prior to and after sintering. As can be seen from the table, as a result of the sintering the carbon content in the specimens is reduced compared to the initial content, and the higher the degree of oxidation of the powders, the greater this reduction.

The relative amount of carbon burnt up during sintering for different degrees of oxidation of the iron powder in the charge is shown in Fig. 3. According to the graph, when the oxygen content is 2%, three times more carbon is burnt up than when it is 0.7%. A reduction in the amount of carbon in the specimens results in an improvement in the frictional properties of the sintered mixture, mainly an improvement in wear, which is reduced from 0.44 to 0.20 mm.

While the use of nonprereduced (oxidized) iron powders in the charge with an increasing amount of oxygen for the production of friction products from sintered mixtures results in a certain re-

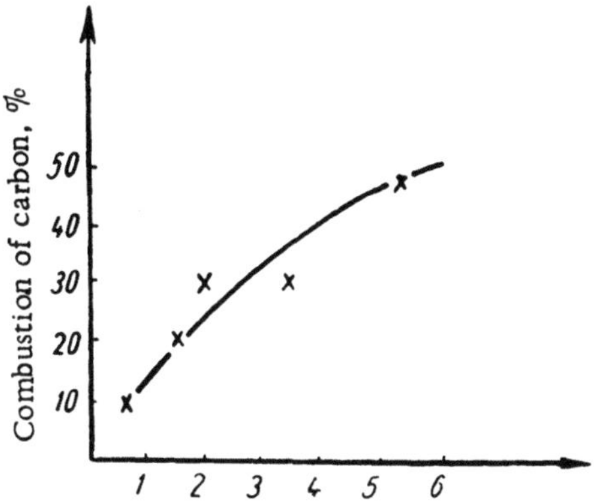

Fig. 3. Relative combustion of carbon as a function of oxygen content in iron powder.

duction in the density of the pressings and poorer pressibility of the powder, the quality of the pressings is still satisfactory. They are not laminated and do not crumble.

The sintering of these parts is characterized by a change in chemical composition. The higher the degree of oxidation of the iron powders, the more the carbon is burnt up. Reduction of oxide films with the formation of active iron atoms improves the contact surface between grains and increases shrinkage during sintering, and this fact, in turn, causes an increase in density. As a result, the density of the sintered parts after sintering proves not to be lower, but even higher, when the degree of oxidation is high, than the density of the ceramic mixture prepared from prereduced powders containing 0.7% oxygen.

The increased porosity and reduced carbon content of the sintered mixture leads to a change in the mechanical and frictional properties. Hardness is increased and wear resistance is improved. The friction coefficient remains virtually constant up to 400°C, and only slightly increases between 300 and 900°C. Minimum wear in ChNMKh pig iron is obtained when it is paired with a sintered mixture containing iron powder with 1.5-2% oxygen in the charge.

Iron powder mark PZh-3 in the as-delivered state contains not more than 2% oxygen. Data on the properties of sintered mix-

Table 4

Properties of sintered mixture with iron base	Powder	
	prereduced	nonprereduced
Oxygen content in iron powder, %	0.7	2.0
Density of pressings, g/cm^3	5.22	5.02
Density of specimens after sintering, g/cm^3	5.6	5.94
Hardness H_{RF}	70	89
Wear of sintered mixture, mm	0.44	0.23
Wear of pig iron, mm	0.33	0.22
Friction coefficient at 1000°C	0.20	0.19
Carbon content, %	5.69	4.50

tures obtained with this powder are shown in Table 4. The table also gives data on the properties of a sintered mixture prepared with prereduced powder. Comparison of the data obtained shows that the properties characterizing the quality of the friction parts – wear-resistance, hardness, and density – are greater in the case of a sintered mixture made with unprereduced powder. The carbon content during sintering is only slightly different and can be easily compensated when making up the charge.

On the basis of our tests it may be considered possible to discontinue the operation of prereducing iron and copper powders when making parts from sintered friction mixtures.

Summary

1. It has been found possible to use nonprereduced iron and copper powders (PZh-3 and PM-2 powders) in the as-delivered state for serial production of parts made from frictional ceramic mixtures with an iron base.

2. The use of these powders simplifies the technological process, reduces the amount of dust in the shop atmosphere, decreases the labor consumption by 10-15%, and ensures the production of high-grade parts from sintered friction mixtures.

References

1. Bal'shin, M. Yu., Powder Metal Science, Metallurgizdat, Moscow, 1948.
2. Bebnev, P. I., in: Powder Metallurgy (Reports of the Fourth All-Union Scientific-Technical Conference on Powder Metallurgy), Yaroslavl', 1956.
3. Fedorchenko, I. M., Poroshkovaya.Metallurgiya, No. 2, p. 27; No. 3, p. 17, 1962.

Chapter 8

Trends in the Development of Bearing and Friction Materials on the Basis of Iron—Graphite—Pores

L. Jenicek, V. Dufek, and Z. Ministr
CVL Mokrat
Prague, CSSR

The iron–graphite system as exemplified by gray cast iron is a commonly used material for one element of both bearings and piston rings, and brake shoes of railroad wheels. In these applications, the main requirements are wear resistance and the stability of the desired coefficient of friction. These requirements are so important that other properties, such as mechanical strength, must be regarded as secondary.

Considerable work on gray cast iron has been performed in many countries in recent years. The processes for improving its quality have been thoroughly investigated, so that a significant improvement of its sliding properties cannot be expected in the near future. On the other hand, powder metallurgy opens new possibilities for the iron–graphite system, as it has done for many other materials.

Porosity is an important variable in the sintered iron–graphite system. The pores are very uniformly distributed and their sizes and volumes can be highly variable. The effect of porosity on sliding or bearing properties can be so pronounced that the knowledge and experiences derived from cast or wrought materials cannot be easily transferred to the development and use of P/M materials [1, 2].

Several metallic additives display a substantial effect on the sintering of iron powder, especially copper. These additives are also important in the manufacture of friction and bearing materials, and two authors of this paper have obtained noteworthy results with the addition of copper, lead, and bronze [3, 4]. However, when a survey of the many examples of the system with various additives is made, the same problem of the manufacturing technology of the basic iron–graphite–pores system is encountered, while the effect of the other additions, important as it may be, is not significant.

Sintered iron–graphite is used for bearings as well as for clutch and brake linings. The actual metallic pairs are usually divided into sliding and rubbing pairs, according to the type of application. When liquid lubricants are added, we talk of lubricated pairs (bearings or clutches); if they are absent, we talk of dry rubbing pairs. Neither of these criteria has a decisive influence on the selection of the material. So-called dry bearings (i.e., sliding pairs) and lubricated clutches (i.e., rubbing pairs) very often have the same coefficient of friction (from 0.1 to 0.15).

The designations "sliding parts" and "rubbing parts" are also used in this paper; however, from the standpoint of the material a distinction is made whether the heat generated during operation is low or high. The operating temperature of lubricated and dry bearings and lubricated clutches is low and does not usually exceed 100°C, unless the actual surface layers are considered. Dry clutches and brake linings operate at higher temperatures. The requirement imposed on highly loaded brakes to withstand short-timed peak temperatures as high as 1100°C is responsible for the greater interest in iron–graphite than in bronze–graphite.

No metallic friction material itself can ensure the stability of the coefficient of friction within a wide temperature range unless a suitable nonmetallic stabilizer is present. Graphite is very suitable for this purpose, and is often used in rubbing pairs operating at low temperatures. When it acts as a solid lubricant, e.g., between the mating surfaces of a bearing and journal, a thin layer of graphite adheres to the journal, yet no chemical reaction or diffusion of carbon into steel occurs.

If the temperature of the rubbing bodies is high, a frictional interlayer is formed due to the melting of the surface asperities of the cast iron element. The graphite present in the lining contributes at least to the formation of a reducing medium, or in other words, it prevents oxidation, which would cause excessive wear.

When two identical brake linings rub against each other, no marked transition layer is formed. This proves that the layers which are created under the surface of ordinary materials, such as steel or cast iron, as a result of chemical reactions, mutual solution of various components, and phase changes play a very important role.

The hardness of the interlayer (transition layer) is important and its existence can be proved by microhardness measurements performed at room temperature. It can easily be deduced that the thickness of the interlayer cannot grow with time of rubbing. If this were the case, it would have to peel off at irregular intervals, thus causing intermittent changes in frictional force. In an ideal situation a dynamic equilibrium should be formed on the boundary, corresponding to the given frictional conditions.

The reproducibility and stability of these interlayers, which secure the minimum wear and the best stability of friction, depends on various conditions which are closely related to the composition of the material, as well as the design of the friction aggregate.

Characteristics of the Basic Components

Atomized iron powder and natural powdered graphite are normally used for the manufacture of iron–graphite sintered materials. Sponge iron powder is just as suitable for materials designed for high temperatures. A pearlitic structure is desired when iron is sintered in the presence of graphite. Its coefficient of friction does not differ from that of other structural components in the iron–carbon system. However, its resistance to wear is superior. Pure graphite is commonly regarded as a sliding (lubricating) component because of its hexagonal structure. However, experimental data indicate that its effect is not so clear cut. Bowden [1] claims that on complete degassing the friction coefficient of

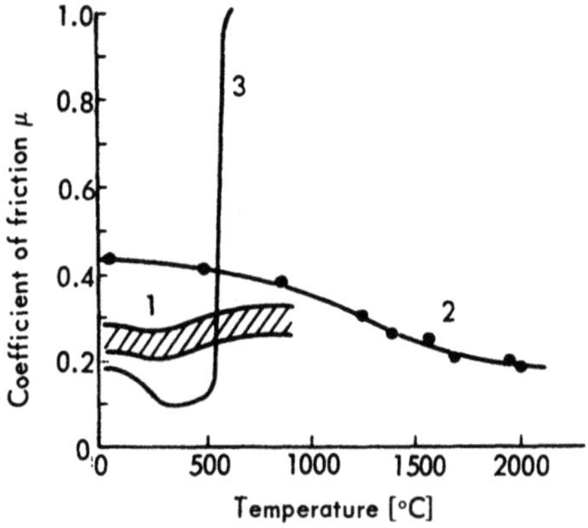

Fig. 1. Coefficient of friction as a function of temperature: 1) according to our experiments on the I-47 testing apparatus; 2) after Kenyon; 3) after Deneon and Goodman.

graphite rises from about 0.1 to 0.6 and the wear increases very significantly at the same time. Yet a fairly low pressure (0.2 torr) of oxygen, water vapor, hydrogen, or hydrocarbons causes a drop in the friction coefficient from 0.48 to 0.15-0.20. We have plotted our results (1 in Fig. 1), and compare them with data derived from experiments performed in a vacuum (curve 2) and data obtained from tests carried out in air with a graphite film applied to a platinum surface with a coefficient of friction of 1.8 (curve 3). The sudden rise in the coefficient of friction in the vicinity of 600°C was accompanied by rapid oxidation, causing a loss of material. In our opinion, the loss of graphite at this temperature would not be accompanied by a rise in the friction coefficient if graphite were impressed into a porous surface. (In fact, the observed rise in the coefficient of friction was merely the return to the coefficient of friction of the platinum itself after the very thin graphite layer had been burned off.) No distinct change was observed above 100°C, which is considered by some (e.g., [5]) as a limiting temperature for the good sliding properties of graphite. However, such a simplified view based on the decisive role of the adsorbed moisture cannot explain the properties of iron–graphite materials which operate at elevated temperatures in air.

We have no experimental data of our own on iron–graphite materials with more than 50% graphite. In Table 1 the results of

Humenik et al. [6] show that an increase in graphite in an iron–graphite bearing material causes an increase in the coefficient of friction and in wear. Similar observations were published by Pugina [7], who claimed to have attained a friction coefficient of 0.40 by raising the graphite content from 50 to 90 vol.%. It must be noted, however, that the value of the friction coefficient determined by laboratory tests is strongly dependent on the type of the testing apparatus.

Sliding Materials with Little Heat Generation. Bearings with Interconnected Pores

Graphite plays an auxiliary but important role in oil-impregnated bearings. Its importance is not yet fully appreciated and so far is not fully explained. Graphite is added to this material to obtain a pearlitic structure after sintering, without free cementite. There is no unanimity in the current literature as to whether and how much free graphite should remain. In our opinion, the presence of free graphite increases the resistance of bearings against scoring, increases their load-carrying capacity, and widens their applicability to both higher and lower temperatures. The amount needed is small (up to 2%) because only a small amount of graphite is required to form an adhesive surface film which can separate the elements of the sliding pair.

Researchers who have investigated the mechanism of lubrication of porous metal bearings claim that oil flows continuously from the pores to the surface and back (see, e.g., [9]). Ministr [9] studied the flow of oil and other liquids through a bearing wall of porous iron. He concluded that the flow rate y was not constant

Table 1. Effect of Graphite Content on the Coefficient of Friction and Wear of Iron–Graphite

No.	Graphite, vol. %	Fe	Coefficient of friction	Average wear, mm/2500 hr
1	100	–	0.13-0.20	–
2	80	20	0.18-0.19	0.042
3	70	30	0.15-0.16	0.030
4	60	40	0.15-0.16	0.010
5	50	50	0.13-0.14	–

and decreased rapidly with time x, approximately as

$$y = a^x,$$

where a is a coefficient expressing the intensity with which the pores are constricted by a pure liquid. The tests indicated that the constriction of pores reaches a stabilized state independently of the effect of the overpressure and also without any flow of the liquid; only the time of contact was decisive.

In the light of this knowledge it cannot be assumed that the flow of oil is essential for attaining sliding properties. However, it is reasonable to increase the additional amount of oil in the bearing or to support the oil flow.

These findings have led to the use of graphite as the main sliding component in the lubrication of porous bearings rather than sole reliance on oil flow. This can be achieved by two basic methods.

Increase of Graphite Additions

The first traditional process is the increase of the graphite content in the iron–graphite system using manufacturing method 1 or 2 in Table 2. A number of individuals recommend this method, although they are aware of the technological difficulties involved, because they do not realize that it is not necessary to have graphite, which should form a component of the sliding surface, distributed in the whole bearing wall. Also, the amount of graphite necessary to obtain the desired sliding properties (resistance to wear and life with regard to the thickness of the sliding layer) has not been determined. Humenik [6] has made considerable progress in this direction in the previously cited work with the graphite cermets which contain up to 80 vol.% graphite. However, the technology is complicated and the material obtained is brittle and difficult to cold-form.

Another alternative is the partial replacement of graphite by lead. In the the German Federal Republic [11-15] and elsewhere [16-18] the manufacturing technology of bearings from iron, lead, and graphite powder and/or molybdenum disulfide has been protected by patents. In some cases, lead is replaced by lead oxide PbO [19-20]. Spreadborough [21] claims that lead oxide is as suitable a dry lubricant as graphite, especially above 200°C, when graphite

Table 2. Metal–Graphite System Manufactured by Powder Metallurgy

Original state	Method of manufacture	Product and its structure
Mixture of iron powder, alloying metals and graphite (or MoS_2); in the case of friction materials, also an addition of nonmetallic friction additives.	1. Compacts are sintered, coined, and oil-impregnated.	Bearing bushes with interconnected pores and other components with ferritic, pearlitic-graphitic structure, and porosity from 10 to 30%
	2. Compacts are recompacted and densified after sintering, or brazed to a steel backing by means of pressure sintering	Bearing and especially friction materials with analogous structure but without interconnected pores (porosity from 5 to 15%)
Iron powder, possibly with addition of other metals. Graphite, MoS_2, or a mixture of them with other metallic or nonmetallic additives is introduced into the sintered skeleton.	3a. Compacts are sintered, coined, and impregnated with colloidal graphite. 3b. Graphite or MoS_2 with other additives is rubbed into the pores of the surface.	Bearing bushes (porosity 10 to 30%, ferritic structure) with interconnected pores, which are filled at least on the surface with a suspension of graphite or other sliding additives.
	4a. Iron powder (possibly in a mixture with other metal powders) is sintered to a steel strip. The pores are subsequently filled with graphite or MoS_2. 4b. A mixture of lead and graphite is rolled or pressed into the pores of the iron–copper bearing skeleton. Lead can be partly replaced by copper or bronze.	A brazed sliding layer, the pores of which are filled with graphite or MoS_2.

fails. However, these bearings have not reached high volume production because of the many difficulties connected with the sintering of compacts made from iron and lead powder or lead oxide and graphite.

Infiltration of Porous Iron with Graphite in Oil Suspension

Graphite in an oil suspension has not been used as a lubricant because it may seal the pores and form a sticky paste. The pores make possible the flow of oil which effects the sliding or bearing properties.

A paste forms, for instance, when carbon particles from electric brushes get into the oil-impregnated iron bearings after several hundred hours of operation. After cooling the paste is so sticky that it glues the journal to the bearing. However, the filling of the pores with pure graphite that contains no tarlike components differs considerably from the filling of pores with worn-off particles, impurities, and residues of depreciated lubricants.

The Research Institute of the Automobile Industry, investigated the problems of lubrication of porous bearings, durability of oil, and the flow of several liquids and oils in porous iron. The investigation was directed at the lubrication of bearings of porous iron containing pure noncolloidal graphite which was introduced into the pores of the surface. The procedure ensured the distribution of graphite in a layer corresponding to the permissible wear of the bearing, and also a suitable orientation of graphite particles.

The difference in the behavior of colloidal and natural graphite upon compression into a porous bearing, is important. Colloidal graphite dispersed in oil can be easily infiltrated through the wall of a porous-iron bearing. It can also be easily and almost completely removed from the pores, by means of hot benzene, even several thousand hours after the infiltration.

On the other hand, when an oil suspension of pure graphite of particle size comparable to the pores is infiltrated, without an antisedimentation agent through a bearing wall of porous iron, graphite penetrates with difficulty but then adheres so strongly to the inner surface of the not-too-deeply situated pores that it cannot be removed by a prolonged immersion in boiling benzene, even if

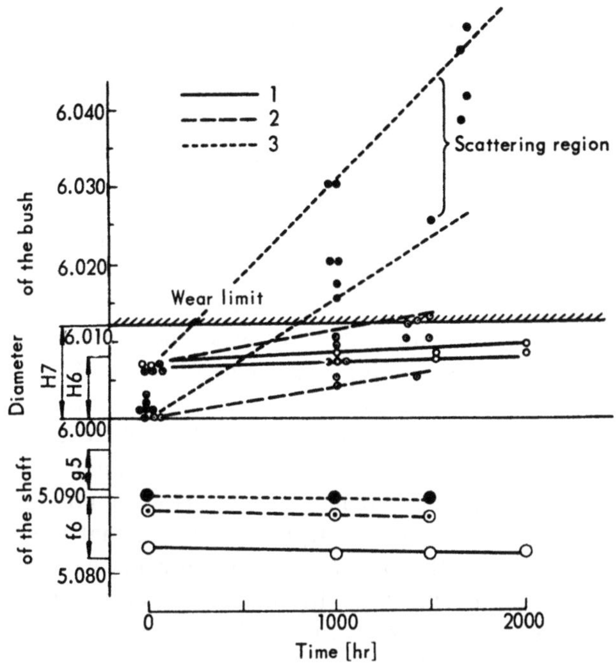

Fig. 2. Wear of bearing from porous iron infiltrated with oil and graphite (1), with oil without graphite (2), and of a DU bushing (3) in an electric motor of an automobile heater (manufactured by PAL, Kbely, 12 V, 2800 rpm). Shaft: plain carbon steel CSN 11600, roughness 0.4μ, load by air brake; atmosphere: flowing dry air; temperature of the surrounding: 20 to 80°C cycled each 5 hr.

the oil–graphite suspension has been in the pores for only a few minutes.

Several methods of introducing graphite into bearing pores and various concentrations of oil–graphite suspension were investigated at the Research Institute of the Automobile Industry. Beside compressing and infiltration due to capillary action the rubbing-in of an oil suspension with 20% graphite by means of a knurled coining punch has proved most successful. Good results were also obtained from rubbing-in oil-suspended graphite with the aid of a cylindrical brush in the already coined bearing aperture.

The amount of graphite (without oil) introduced into the bearing surface by means of the knurled coining punch is approximately

2 mg/cm^2. The sliding surface turns clearly dark and the pores become filled with graphite to a depth of several hundredths of a millimeter. This considerably improves the sliding properties and the life of the bearing (and the life of the punch as well). Although the amount of oil introduced into the bearing is very small, it safely protects the porous iron against corrosion.

The endurance of these bearings was measured in electric motors of automobile heaters, which were tested on a universal test apparatus for electric rotating motors. Motors with porous-iron bearings with oil and without graphite and motors with Glacier DU dry bearings were also tested. The tolerance H7 was selected as the permissible wear limit. The results, given in Fig. 2, clearly show a wide range of variations in wear. The DU bearing wore the fastest and became very noisy after only a few hundred hours of operation (curves 3). The porous-iron bearing lubricated with oil without graphite wore so much after 1500 hours that the bearing aperture exceeded the H7 tolerance limit (curves 2). The average wear of the bearing lubricated with oil and graphite was only 1-2 μ after 1500 hours of operation (curves 1). In this case, only the lower and upper limit values are plotted in Fig. 2, instead of all six test results.

The wear of the journals was in all cases very small and hardly measurable. In the oil-and-graphite-lubricated bearing the graphite adhered to the journal and it was not practical to wipe it off during the measurements. Brush wear was very small in motors equipped with DU bearings or oil-and-graphite-lubricated bearings, although none had oil slingers. In motors with oil-lubricated bearings, the commutator could not be protected from contamination with oil, even by means of an oil slinger. Excessive wearing of the carbon brushes resulted.

When the motors were tested at a low temperature of −40°C, those with bearings lubricated with oil and graphite and the DU bushes exhibited quick-starting and noiseless operation. The motors with porous-iron bearings lubricated only by oil started slowly only when the input was increased and often produced a shrill squeak which disappeared only after the bearings had warmed up.

Loading tests showed that bearings with graphite (20 mm in diameter) can carry as much as 240 kp/cm^2 at a velocity of 0.1 m/sec without any signs of seizure. The endurance of bearings containing graphite was 50 to 100% higher than that of graphite-free bearings and, at the same time, their coefficient of friction was lower.

The simple coining of the bearing aperture connected with the rubbing-in of the graphite paste can diminish the deviation of the sliding surface from cylindrical form, which prolongs life and prevents noise.

The technology developed has shown how effective even a small addition of graphite can be when it is fully exploited. The bearings operate reliably even at temperatures at which ordinary oil-impregnated bearings fail.

Graphite-impregnated bearings must have a minimum wall thickness of 2 mm with respect to compacting parameters, which can limit their application. For example, they are not suitable for use in automobile chassis.

Bearings of Iron–Copper–Lead Alloy with Graphite Addition

In order to eliminate the above-mentioned disadvantage and to lower the price of nonlubricated bushes, Fe–Cu–Pb alloy bearings have been manufactured for a number of years from a steel strip lined with a brazed metal powder together with a nonmetallic solid lubricant. The most widely used are Glacier DU bushes.

Favorable experiences were obtained with iron–graphite bushes which contain graphite only in the surface layer. Because of this Ministr conceived the idea of applying by rolling a surface layer of a mixture of graphite and lead, and possibly copper, to a sintered iron–copper layer (see Table 2). He produced a new type of bearing bush with a thin metal–graphite layer by this procedure. A 0.2-mm layer of a mixture of iron and copper powder is sintered on a steel strip; this is followed by rolling a mixture of copper–lead and graphite powder, which is impressed into the voids of the sintered layer. The strip is then cut and bent into bushes. The composition of the sliding layer is (in wt.%) 26 Fe + 27 Cu + 45 Pb + 2 graphite (i.e., 8 vol.%).

In contrast to the Glacier DU bushes, in which the bearing layer consists of porous metal (porosity 35–40%) filled with a mixture of lead and PTFE, Ministr's bushes have a bearing layer consisting of a monolayer of brazed metal powder particles that have voids filled with copper–lead and graphite. This pattern allows an

Table 3. Life of Various Bearing Bushes in the Starter of a Skoda 1000 MB Automobile

Material	Number of starts
Thick-walled bearing bushes:	
(a) from porous iron with 0.1% C and 25% porosity (infiltrated with oil)	5000
(b) from a porous Fe–Cu–graphite material with 20% pores (infiltrated with oil)	10000
Double-layer material with a sliding layer:	
(a) porous bronze with 30% of pores (filled with a mixture of 1 part of lead and 4 parts of PTFE)	7000
(b) Fe–Cu–Pb–graphite	25000

increase in the portion of sliding components from 35 to 65%. It is most important to minimize the resistance against their impressing into the bearing matrix. In this way it is possible to produce various porous sliding layers, as the resistance encountered, for example, by pressing lead powder into the pores of sintered iron is smaller than the pressure necessary to achieve a total densification of the lead powder.

Bearing bushes with a thin metal–graphite sliding layer with 2% graphite can be favorable compared to DU bushes with a sliding layer of bronze PTFE, and lead. The former display a comparable load-carrying capacity, but only half as large a coefficient of friction and only a fraction of the wear of the latter. The results of tests in automobile starters are given in Table 3. However, it still remains to established whether the selected composition, especially the graphite content, represents the optimum for the given application.

It seemed desirable also to test the metal–graphite layers on clutch plates. Sliding layers with a constant ratio of iron, copper, and lead, and with various contents of graphite (i.e., 0, 1.5, and 3%) were tested. The results of the tests are given in Fig. 3, where data on ordinary porous iron and porous bronze with 1.5% graphite are added for comparison. In Fig. 3c, the change with time of the coefficient of friction and the temperature of specimens containing 1.5% graphite is plotted. This example shows that the determination of the friction coefficient on the test apparatus I-47 has an orientation value only, especially at low specific pressures.

The experiments have proved that a definite minimum addition of graphite is necessary to lower the wear of the material. We consider it necessary to perform extensive testing as to how far the graphite content can be decreased in the above-described bushes, where the bending of metal–graphite layers causes certain difficulties. The positive effect of heating the specimens to 300°C should also be studied (see Fig. 3.)

It is anticipated that this application of graphite, and materials with similar properties, offers many new P/M applications in the field of nonlubricated bearings and other sliding or friction elements such as clutch plates, where the permitted wear does not exceed 0.1 to 0.2 mm. The unique aspect of this concept is the graphite being introduced into the system after the metallic matrix of load-carrying and heat-abducting components has been completed by sintering at high temperatures. We believe that this method of preparing the iron–graphite system is preferable because of the superior sliding properties. We stress the positive influence of the orientation achieved by either rubbing-in or rolling-in of graphite, and also the fact that graphite introduced into an open matrix can more easily form a continuous uniform layer. Thus we can explain the excellent results achieved with this type of bearing material in spite of its low graphite content.

Materials Displaying Considerable Heat Generation

Brake linings operating at temperatures exceeding the resistance of bronze–graphite linings generate a considerable amount of heat. According to Dufek, the optimum particle size is closely related to the required porosity of the product. If the porosity is between 10 to 20%, which seems to be most advantageous for brake linings, the average particle size should be from 50 to 100 μ. Depending on the type of powders used and the amount of additives, it may be necessary to remove the fine and coarse portions. However, particle size is not the only factor determining the final porosity. As can be seen from Fig. 4, the porosity of an iron–graphite brake lining manufactured from an unclassified sponge iron powder can be varied by the choice of compacting and sintering pressure. Free sintering, without applied pressure causes an in-

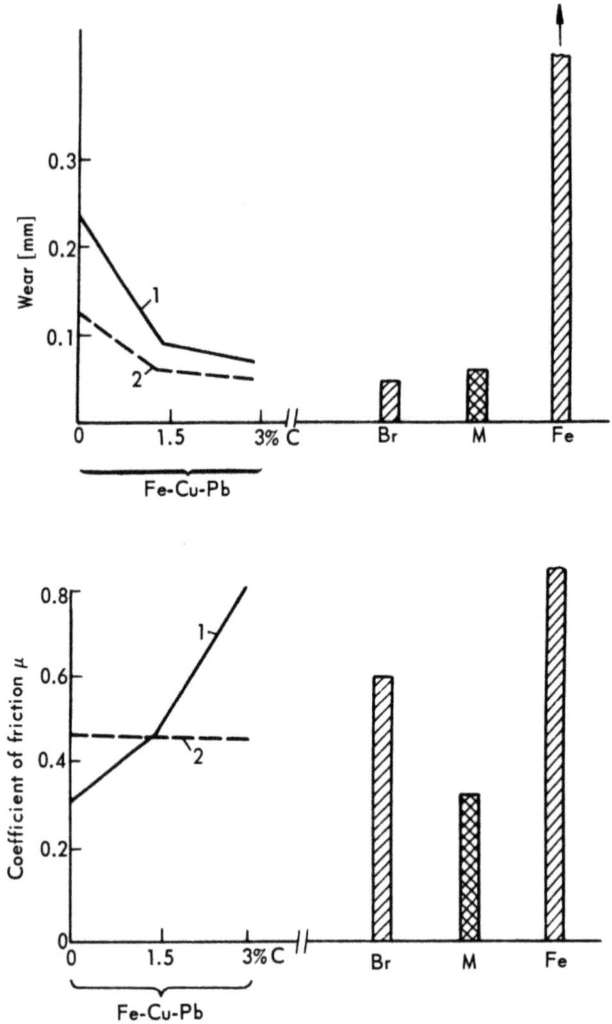

Fig. 3a, b. Wear (a) and coefficient of friction (b) tested on the I-47 apparatus. 1, 2) Effect of the graphite content on the behavior of the lead layer which has been pressed into the iron–copper porous skeleton (1), or subsequently heated at 300°C for 15 min (2). Mating surface: plain carbon steel CSN 11600; temperature of test 220-250°C. Br) Porous bronze with 1.5% graphite, dry, against steel CSN 11600. M) Mintex (brake lining of English origin) dry, against grey cast iron CNMCh. Fe) Porous iron, dry, against steel CSN 11600.

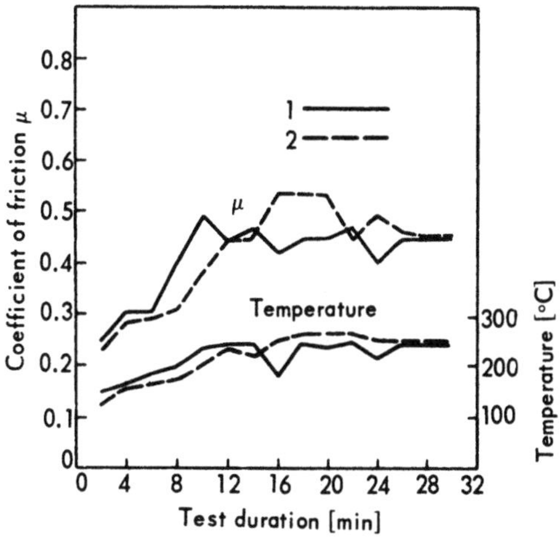

Fig. 3c. Variation of the coefficient of friction and temperature with Fe–Cu–Pb containing 1.5% graphite. Load 5 kp/cm^2. Diameter of the ring-shaped specimen 28/20 mm, corresponding circumferential velocity at 100 rpm 1.465/1.025 m/sec. Testing time 30 min.

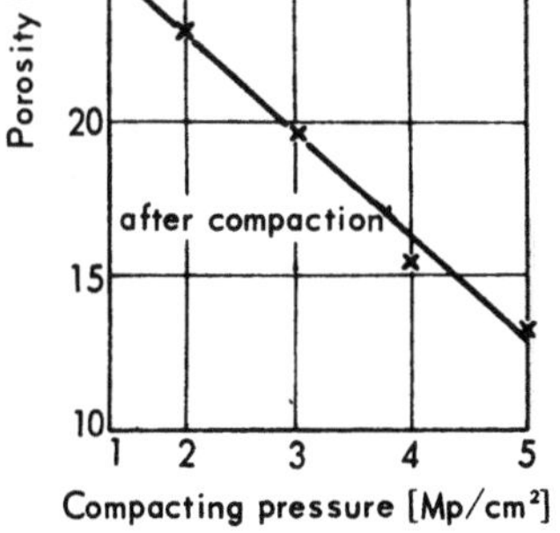

Fig. 4. Effect of compacting pressure on the porosity of iron–graphite friction material Diafrikt F1. (The porosity remains unchanged by sintering under pressure of about 15 kp/cm^2).

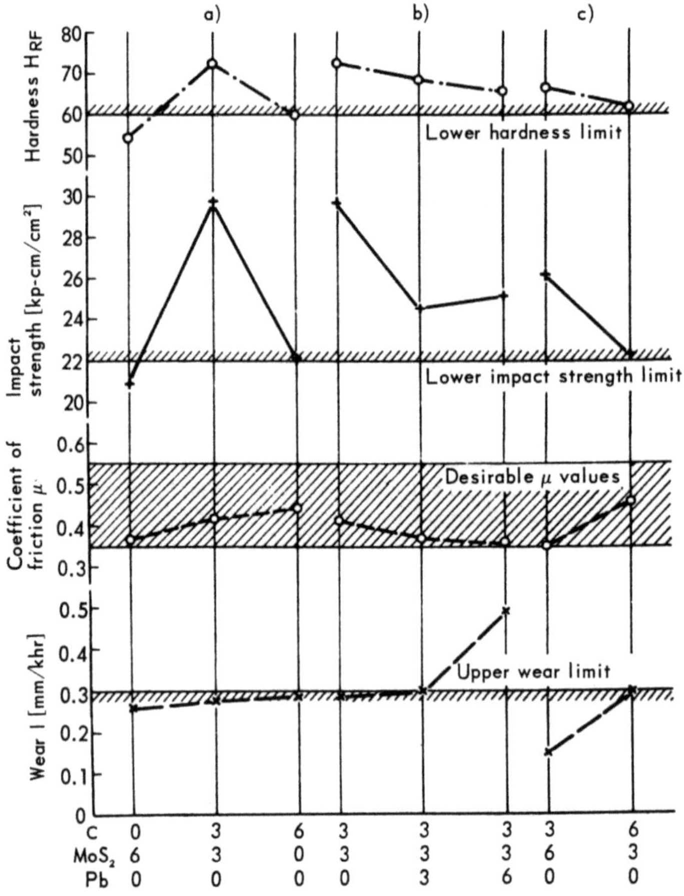

Fig. 5. Effect of the addition of graphite, MoS_2, and lead on the properties of brake lining Diafrikt F1: a) graphite replaced by MoS_2; b) effect of lead addition; c) effect of a simultaneous addition of graphite and MoS_2.

crease in the porosity of the compacts. Sintering under a specific load of 15 kp/cm^2 leads to a porosity roughly equal to the green porosity. Different particle sizes and compacting pressures can be used for certain applications, but the general rule remains valid.

The above stated, almost contradictory observations of the behavior of graphite have led us to the idea of seeking the minimum permissible amount of graphite, instead of the maximum one as other authors have done. Eisenkolb and his co-workers [22] found 21 wt.% (48 vol.%) of graphite as the optimum amount for friction materials. In our opinion, this amount should represent all non-metallic additives necessary to ensure the proper wear resistance of the iron–graphite–pores system; however, it is neither useful nor desirable to fill this portion with graphite only.

Results obtained by Dufek will be discussed to support this assumption. He replaced graphite with molybdenum disulfide in steps and found that a mixture of 3% graphite and 3% MoS_2 displayed the best mechanical properties (see Fig. 5a). To increase the total portion of sliding additives, he investigated the effect of lead additions to this combination (Fig. 5b). However, a decrease in mechanical properties resulted, and wear increased at higher amounts of lead. He then tried to determine whether it was more important to keep the portion of graphite or that of molybdenum disulfide constant, provided that the amount of lead equaled zero (Fig. 5c). The answer is still uncertain because 3% increase in

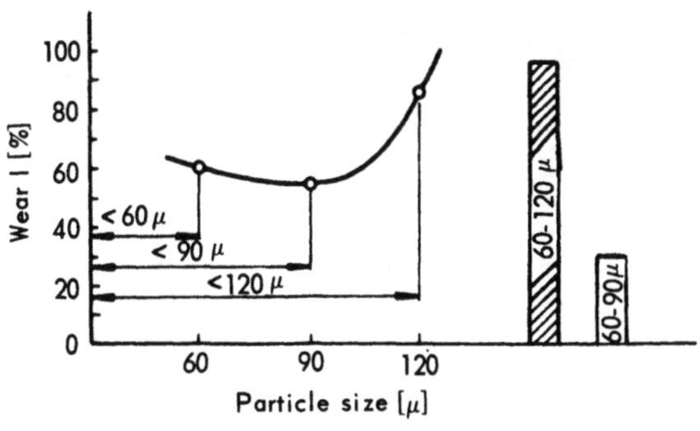

Fig. 6. Effect of particle size of graphite on the wear of the iron–graphite friction material (Diafrikt F1). (Permissible wear 1 mm per cycle is plotted as 100%.)

Table 4. Carbon Content in an Iron–Graphite Friction Material during Sintering in a Nitrogen Atmosphere.

No.	Binding component*	Graphite addition %	Actual carbon content, %†	Carbon content after sintering			Loss during sintering, %	Related to total carbon in the mixture, %
				C_{total}	C_{free}	C_{bond}		
1	A	13.4	11.24	9.75	8.77	0.98	1.49	13.2
2	B	9.0	8.1	6.91	5.84	1.07	1.19	16.6
3	A	9.0	8.1	7.25	6.25	1.00	0.85	10.5
4	B	6.5	5.85	3.73	not determined		2.12	36.0
5	A	6.5	5.85	4.15	"	"	1.70	29.1
6	A‡	6.5	5.85	4.73	"	"	1.02	17.5

* A) bronze; B) copper.

†Since the graphite contains 10% impurities, the actual carbon content is given in the second column.

‡Sintered in hydrogen.

Note: In specimens 3 to 6, the relative loss was more than 15% (related to the carbon content in the original mixture). They contained 6% $BaSO_4$, which theoretically consumes 0.62% carbon for the reaction $BaSO_4 + 2C = BaS + 2CO_2$.

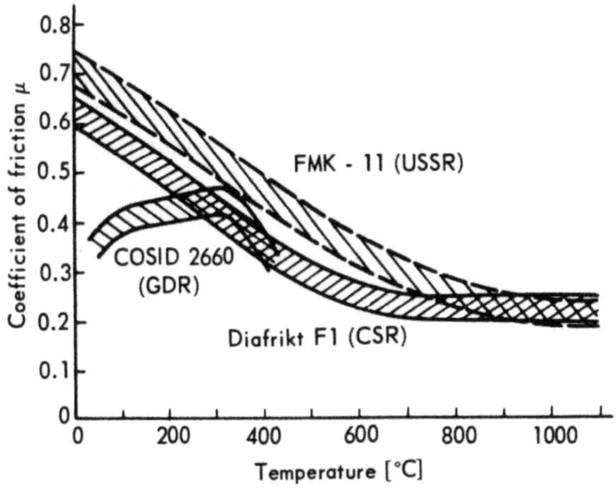

Fig. 7. Effect of temperature on the coefficient of friction of different brake linings.

MoS_2 led to the lowest wear values and an unwanted reduction of the coefficient of friction. The strength and hardness remained above the tolerable limit. The increase in graphite by an additional 3% confirmed the experience that not only wear but also coefficient of friction can be raised by this technique.

Dufek has also studied the size of graphite particles, added in the portions from 5 to 10%. The natural graphite used contained 8 to 15% of impurites, mostly iron oxide and silicon dioxide. These impurities do not lower the quality of the friction material.

The most important conclusions of this work are summarized in Fig. 6. It can be seen that the finest graphite particles impair the wear properties just as much as the coarser ones. The best results were obtained with the fraction 60 to 90 μ, but economic considerations have led to the use of the size range of 25 to 90 μ. Table 4 shows the results of chemical analyses of specimens of iron–graphite friction materials with various contents of graphite and binders. On the basis of these results we can claim that 1% carbon is needed for the formation of a pearlitic structure. The remaining roughly 2% of free graphite are enough to ensure the functioning of the friction lining, if further additions such as $BaSO_4$, MoS_2, or possibly lead are present. Their selection should be based on which one of them lowers the mechanical properties to a lesser extent than graphite.

A comparison of the properties of iron–graphite friction materials produced by different manufacturers is shown in Fig. 7. It can be seen that a high graphite content, as represented by the East German material Cosid, has the advantage that its coefficient of friction is almost constant in the temperature range of 100 to 400°C. But it cannot compete with the Soviet FMK-11 or the Czechoslovak Diafrikt F1 under more difficult operating conditions. On the other hand, these materials display an undesirably high coefficient of friction at low temperatures, at least for certain applications.

Conclusion

Graphite in precisely controlled amounts can be beneficial in P/M iron–graphite systems, along with other additives in applications requiring either low or high temperatures of friction. For materials designed for low temperature it must be induced after sintering–after the porous metallic skeleton has been formed. Materials used for high-temperature friction must receive the graphite by mixing before compacting.

References

1. Bowden F. P., and Tabor, D., The Friction and Lubrication of Solids, Clarendon Press, Oxford, 1964.
2. Kragelskij, I. V., Friction and Wear, Moscow, Mashgiz, 1962.
3. Dufek, V., Sintered Friction Materials, SNTL, Prague, 1965.
4. Ministr, Z., and Priester, J., Proc. II. Int. Powd. Metall. Conf. Tatra, 1966, Vol. III. p. 71.
5. Prochazka, V., and Miskovic, V., Kovove materialy II, 4, Bratislava, 1964, p. 408.
6. Humenik, J. J., Hall, D. W., and Van Alsten, R. L., Metal Progress, Vol. 81, No. 4, p. 101, 1962.
7. Pugina, L. I., Fedorchenko, I. M., and Ponomarenko, N. E., this volume, Chapter 5.
8. Cegielski, W., and Rutkowski, W., Sintered Bearings, PWT, Warsaw, 1960.
9. Ministr, Z., Strojirenska vyroba No. 12, p. 725, 1964.
10. Designer's Manual, The Glacier Metal Company Limited Alperton Wembley, England.
11. GFR Patent 714 396.
12. GFR Patent 861 700.
13. GFR Patent 891 279.
14. GFR Patent 945 565.
15. GDR Patent 22 323.
16. British Patent 719 146.

17. British Patent 807 419.
18. Japanese Patent 5 352.
19. GFR Patent 895 002.
20. French Patent 883 407
21. Spreadborough, J., Wear Vol. 5, No. 1, p. 82, 1962.
22. Strobel, E., Die Technik No. 12, p. 819, 1957.

Chapter 9

Iron-Based Sintered Antifriction Materials for Heavy-Duty Service

I. M. Fedorchenko, L. I. Pugina, V. V. Pushkarev, J. F. Shevchuk, and V. S. Ageyeva

Academy of Sciences of the Ukrainian SSR
Kiev, USSR

The development of new equipment and machines for heavy-duty service demands the development of new bearing materials capable of service at high loads and sliding speeds. One of the types of service of highly loaded friction elements is the application as a sliding pair of raw, thermally nontreated metallic surfaces. Another type of application is service under conditions of boundary lubrication. This mode of operation is the result of increased loads and sliding velocities.

Present sintered iron–graphite materials are able to serve up to loads of 20 to 30 kp/cm^2 and velocities up to 1 m/sec. Sulfidation and alloying of iron–graphite, together with proper heat treatment of the counterface material, can increase the allowed service parameters by an average factor of two. It is urgent that new antifriction materials be developed capable of supporting unhardened journals at loads exceeding 100 kp/cm^2 and velocities up to 4.5 m/sec under boundary-lubrication conditions. Such conditions of friction require that the material posess a number of

Table 1

No.	Marking of material	Porosity, %	Hardness HB, kp/mm^2	Cut-off spec. load, v = 4.5 m/sec	Coefficient of friction	Time for adaptation, min
1	Fe ZnS4 F	16.1	76.3	24	0.01	10
2	Fe G1ZnS4 PF	18.0	47.5	64	0.01	10
3	Fe G3ZnS4 PF	14.6	62.4	52	0.007	20
4	Fe G5ZnS4 PF	21.9	33.1	30	0.015	25
5	Fe ZnS4 PF	15.5	114.0	16	0.02	40
6	Fe G1ZnS4 P	18.7	95.0	52	0.01	90
7	Fe G3ZnS4 P*	13.5	107.0	90	0.007	140
8	Fe G5ZnS4 P*	19.0	84.9	30	0.015	130

*The structure of the material contains up to 5% cementite in the form of a discontinuous network.
Marking: Fe, iron; G, graphite; Cr, chromium; Sn, tin; Pb, lead; Mn, manganese; S, sulfur; FeS, ferrous sulfide; ZnS, zinc sulfide. Numbers following the symbols indicate percentage. P, pearlitic structure; F, ferritic structure; PF, pearlitic-ferritic structure.

additional properties, namely, good adaptability, low coefficient of friction, and resistance against seizure.

The investigation in the field of sulfided iron–graphite materials and the study of their properties have shown that materials containing sulfides of zinc possess the best qualities [1-5].

Analysis of the results of investigations leads to the conclusion that materials with ferritic structure can be used for sliding in difficult conditions of work for the sake of diminishing the wear of the counterface. However, to prevent seizure, the bearing material must contain an increased amount of components acting as dry lubricants (graphite, sulfides, some soft metals, e.g. lead) and being able to prevent the seizure of rubbing surfaces. For the same purpose the matrix material (ferrite) has to be strengthened by alloying elements such as carbon, chromium, manganese, tin, etc.

The objective of this work is to study the effect of the structure of sulfided iron–graphite and of the alloying additions on its antifrictional properties. A selection of materials with optimum compositions was prepared and the effect of different manufacturing variables on their properties tested, with special attention

Table 2

No.	Marking of material		Porosity, %	Hardness HB, kp/mm^2	Cut-off spec. load, v = 4.5 m/sec	Coefficient of friction
1	Fe G1,5 ZnS4	PF	20	40.9†	50	0.006
2	Fe G1,5 ZnS8	PF	19.4	41-47	40-70	0.005
3	Fe G1,5 ZnS8	P	17.8	76.3	30-100	0.01
4	Fe G0,7 FeS4 ZnS4	PF	17.5	40-56	30-70	0.01
5	Fe G0,7 FeS4 ZnS4	P	19.0	76.3	15-60	0.007

†Shearing strength = 6.9 kp/mm^2.

given to resistance against seizure. The selection of test materials can be divided in three groups:

1. Materials containing sulfiding components (Table 1).

2. Materials containing graphite and with various types of structure (Table 2).

3. Materials containing alloying elements (Table 3).

The test specimens were prepared by powder-metallurgy methods. The powders were blended for 4 to 6 hr. The compacting

Table 3

No.	Marking of material	Porosity, %	Hardness HB, kp/mm^2	Shearing strength, kp/mm	Cut-off spec. load, v = 4.5 m/sec.	Coefficient of friction
1	Fe G1,5 ZnS4 Cr1	19.5	76.3	18.3	16-20	0.04
2	Fe G1,5 ZnS4 Cr3	19.0	107	23.0	12	0.03
3	Fe G1,5 ZnS4 Pb2	19.3	61.5	16.4	32-44	0.09
4	Fe G1,5 ZnS4 Pb4	19.0	64.0	14.0	44-52	0.009
5	Fe G1,5 ZnS4 Pb6	18.5	60.0	12.8	8-16	0.07
6	Fe G1,5 ZnS4 Mn6 S1	22.7	47.5	7.8	60	0.005
7	Fe G1,5 ZnS4 Mn3 S0,5	22.5	49.1	6.9	40	0.02
8	Fe G1,5 ZnS4 Sn0,5 S0,5	19.2	43.7	18.9	48-100	0.004
9	Fe G1,5 ZnS4 Sn0,5 S1	19.7	51.9	17.1	76	0.006
10	Fe G1,5 ZnS4 Sn0,5 S2	23.4	47.5	17.1	100	0.006
11	Fe G1,5 ZnS4 Sn1,0 S0,5	23.0	43.7	16.9	100-56	0.004
12	Fe G1,5 ZnS4 Sn1,0 S1,0	23.2	47.5	13.8	100	0.006
13	Fe G1,5 ZnS4 Sn1,0 S2,0	19.2	47.5	16.2	40-56	0.006

pressures varied from 5 to 8 Mp/cm^2 according to the composition of the compact. The sintering temperature was 1050° to 1100°C according to the composition and desired structure. The specimens were sintered for 2 hr in hydrogen, using a powder fill to control the resulting type of structure.

Friction properties were measured on a test rig where three bushes were tested simultaneously. The journals (counterface) were made from normalized shaft steel containing 0.34% carbon. Their surface finish was class 8. (according of GOST 2789-59).

The cut-off load (on the verge of seizure) was determined at a sliding velocity of 4.5 m/sec. The load was increased by 8 kp additions for 10,000 to 40,000 revolutions after the stabilization of the frictional moment. The onset of seizure was observed by a rapid rise in the moment of friction.

The coefficient of friction and the adaptability of the specimens were tested at a velocity of 0.9 m/sec, load 50 kp/cm^2, and boundary lubrication. For the purpose of lubrication, "Veretennoje 3" oil was dropped at 15 drops per minute into the rubbing zone.

The experiments have proved that the change in the content of free graphite in the material has a more important effect on the cut-off load than the structure of the material (Table 1). A ferritic-pearlitic material displays maximum cut-off load at 1% content of free graphite, whereas a pearlitic material has a maximum only at 3%. This can be explained by the fact that graphite, besides improving antifriction properties, also increases the strength characteristics of active surface films.

More than 5% graphite lowers the strength properties, which brings about a decrease in the cut-off load before seizure occurs (Table 1). Graphite-free material, even if it turns completely pearlitic during sintering (owing to the carburizing effect of the filling), also has poor resistance against seizure. Stronger materials with pearlitic structure are less adaptable. The coefficient of friction is much the same for all structures (except for materials compacted from graphite-free mixtures) and equals 0.015-0.007. It is now evident that high cut-off loads can be attained both with ferritic-pearlitic and pearlitic materials by controlling the amounts of free graphite.

The increase in the content of sulfides in the composition does not guarantee stability of either the structure or the properties of the material. The strength decreases when the content of sulfur exceeds 1% [1, 2]. This also affects the antifriction properties. The cut-off load varies from test to test (Table 2). This may be caused by brittle fracture of different structural components having various volume stresses. Another factor may be the damage of the surface due to smearing conglomerates whose boundaries were weakened by the presence of a sulfide film. Alloying with chromium increases the hardness of ferrite up to 150-170 kp/mm^2, which leads to a low specific cut-off load (Table 3). On the sliding surface of these materials phase transformation occurs due to high local pressures and high temperatures. A film of secondary austenite is formed, possessing a lattice with the parameter $a = 3.593$ Å, which excludes the formation of stable protective films on the points of true contact.

Similar phenomena occur when manganese is used as the alloying element. Particles of manganese which have not dissolved during sintering are very hard (up to 970 kp/mm^2). The sulfides tend to cluster around them, lowering the strength properties of the alloy and, consequently, its antifriction properties.

The addition of lead did not stabilize the properties and structure of the investigated material. It caused a scattering of the values of the cut-off specific load. A material containing 4% lead and having a pearlite-ferritic structure gave good results when sliding on a counterface of normalized steel. When the content of lead is raised to 6%, the portion of ferrite in the structure increases, which deteriorates the antifriction properties of the alloy. When the content of lead is lowered (up to 2%), it tends to form a pearlitic structure with soft plastic pearlite [1].

Alloying of iron–graphite and sulfided iron–graphite with tin has proved very effective [1, 2, 5]. A similar effect was obtained on all investigated alloys containing tin and sulfur from 0.5 to 2%, in different proportions. All alloys displayed good resistance against seizure. This is explained by a strengthening of the ferrite due to the solid solution of tin in iron, and also by a certain strengthening of surface films in the course of friction. The microhardness of ferrite in the surface film increases from 85-105 kp/mm^2 to 172-203 kp/mm^2 (after friction under 100 kp/cm^2 specific load). At the same time it was noted that both sliding surfaces remained surprisingly clean. An alloy containing 1% tin and 1% sulfur displayed a remarkably high stability; its surface films strengthened during friction at a load of 100 kp/cm^2 from 170-190 kp/mm^2 up to 240-250 kp/mm^2.

Conclusions

Friction experiments undertaken with sintered antifriction materials sliding on normalized steel and designed for heavy-duty service have revealed the following conclusions:

1. To increase the specific load-carrying capacity, the structure must contain free graphite and sulfides to assure the formation of stable surface films.

2. Ferritic-pearlitic structure containing up to 50% pearlite seems to have the best properties.

3. The addition of tin to iron–graphite has a positive effect on its resistance against seizure; when elementary sulfur is added at the same time, α-solid solutions are formed.

References

1. Pugina, L. I., and Shamray, F. I., Powder Metallurgy in Machine and Tool Construction, Kiev, 1961.
2. Pugina, L. I., and Shamray, F. I., Scientific, Technical, and Manufacturing Practice, TsITEIN, 1960, No. M-60-275/9.
3. Fedorchenko, I. M., Filatova, N. A., and Pushkarev, V. V., Fiziko-Khimicheskaya mekhanika materialov, No. 5., p. 567-570, 1965.
4. Fedorchenko, I. M., Drajgov, D. A. and Mamykin, E. T., Fiziko-Khimicheskaya mekhanika materialov, No. 6, p. 670-674, 1965.
5. Pugina, L. I., Second Powder Metallurgy Conference, Cracow, 1963, p. 75.

Chapter 10

Some Recent Advances in the Manufacture of the Friction Sintered Material Grade FMK-11

B. T. Krysin, L. P. Lebedeva, L. N. Ignatov,
and J. V. Kolpakov

A commercial iron-base P/M friction material FMK-11 is being used in aircraft brakes, under conditions of dry friction.

The manufacturing of P/M products made from the FMK-11 material includes the following operations: preparation of powdered mixture, compacting, sintering under applied pressure in hydrogen, machining, and assembling.

The quality of finished products, and their physicomechanical and frictional properties depend on a number of factors, especially the properties of the parent powders and the density of the final products, which in turn, is controlled by the variables of pressing and sintering.

With the aim of improving the quality of current production, simplifying the technology of manufacture, and improving the working conditions, the following investigation programs have been undertaken:

1. The effect of the state of parent materials and technological regimes concerning the pressing and sintering operations on physicomechanical and frictional properties of the FMK-11 sintered material.
2. The selection of an antiadhesive paint to cover interlaying discs used for sintering.

The Effect of the State of Parent Powders on the Mechanical and Frictional Properties of Sintered Products

Iron and copper powders, as delivered, contain a certain amount of oxides in the form of surface films or isolated inclusions; for this reason they are subject to final reduction prior to use. As a rule, final reduction is performed in order to improve some technological properties of the powders, such as their compressibility.

If incompletely reduced powders containing over 1% oxygen are used, it becomes difficult to obtain compacts of complicated shape. At the same time, however, the surface oxides intensify the process of sintering and improve such mechanical properties of sintered products as tensile strength, elongation, and impact strength.

Considering that the products from FMK-11 possess a simple geometric form – plates from 3 to 7 mm thick – it has been suggested that the use of powders as delivered should not make the pressing operation particularly difficult and the process of final reduction could be omitted.

An investigation established the effect of the degree of oxidation of parent powders on the properties of FMK-11. Iron and copper powders, as delivered, were analyzed for oxygen. The results are given in Table 1. It can be seen that the content of oxygen varies in iron powders from 1.10 to 1.66%, and in copper powders from 0.11 to 0.20%. For the purpose of this investigation, iron powders containing 0.7, 1.5, 2.0, 3.5, and 5.4% oxygen were selected. The last three samples were obtained by oxidizing the second sample, containing 1.5% oxygen as delivered. The same was used to prepare the 0.7% oxygen powder by means

Table 1

Type of powder		Oxygen content, wt.%
Iron powder	Sample 1	1.30
	Sample 2	1.66
	Sample 3	1.10
	Sample 4	1.37
	Sample 5	1.54
Copper powder	Sample 1	0.20
	Sample 2	0.11
	Sample 3	0.13
	Sample 4	0.14
	Sample 5	0.15

of final reduction at 600-650°C for 3 hours. As copper powder contains substantially less oxygen than iron powder, the sample with the highest content of 0.20% oxygen was used.

Specimens of the same size as commercial products were prepared for the testing of mechanical and frictional properties. The pressing was performed on a hydraulic press, the specific pressure being 6 Mp/cm^2. The compressibility of the specimens was satisfactory. No separation of layers, cracks, or fissures were observed.

The green and sintered density and the shrinkage of the compacts plotted against the degree of parent powder oxidation, agree with general rules claimed by other investigators. As can be seen from the diagram in Fig. 1, the compressibility of iron powder decreases a little with increasing oxygen content; the green density drops from 5.22 g/cm^3 to 4.93 g/cm^3 as the oxygen content increases from 0.7 to 5.4%; the sintered density and shrinkage increase at the same time. However, when the content of oxygen does not exceed 2%, which is usually true with as-delivered iron powder, the density of sintered compacts does not differ much from that of compacts prepared from finally reduced powders with approximately 0.7% oxygen. The total relative decrease in green density is 4%, the total relative increase in sintered density is 5% in the range of the investigated oxygen content.

The frictional properties were measured on the test apparatus I-47-K-54 using the method developed at Institute of Machine

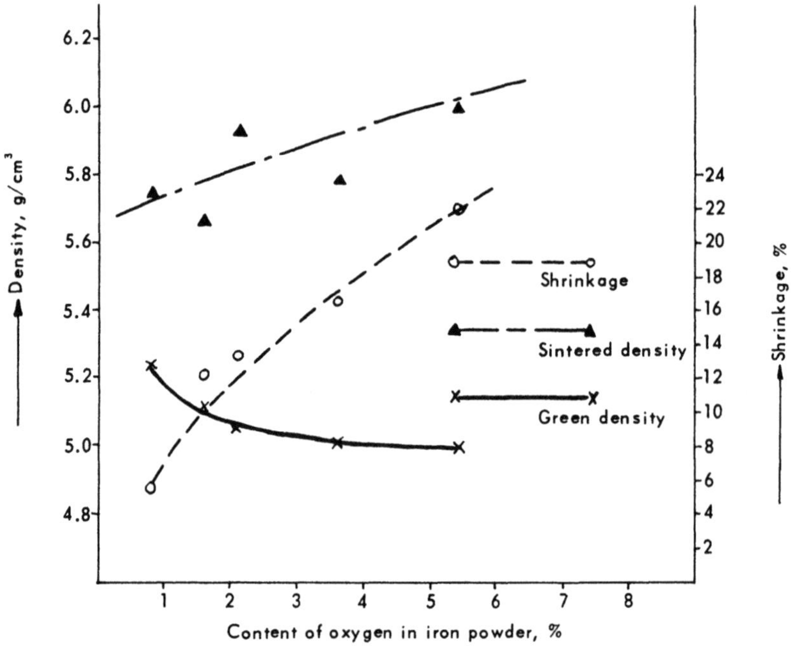

Fig. 1

Science. The sintered specimens were tested by rubbing against specimens made from cast iron grade CNMCH, having a Brinell hardness of 160-220 kp/mm^2. The tests indicated the linear wear related to one complete test cycle and the coefficient of friction at various temperatures. The wear of sintered and cast iron specimens was dependent on the content of oxygen in the parent iron powder. As the oxygen content increased from 0.7 to 5.4%, the wear of the P/M specimens gradually decreased to approximately one-half of the original value. The wear of the cast iron counterbody decreased with increase in oxygen content from 0.7 to 2.0%, then rose again, attaining close to its original value when oxygen reached its maximum value of 5.4%. This means that the wear of the cast iron counterbody is lowest when the oxygen content in the parent iron powder is about 2%, which roughly corresponds to the as-delivered state.

The results of the measurements of the coefficient of friction are given in Table 2, which shows that the coefficient of friction increases a little with increasing oxygen content at higher temperatures (400–900°C).

During the sintering of specimens made from powders with different oxygen contents, a change in chemical composition takes place, namely, the loss or burn-up of carbon, which increases with the degree of oxidation of parent powders. The experimental results showed that when iron powder with 5.4% oxygen was used, the final carbon content was approximately only 47% of the initial amount. The lowering of carbon content apparently improves the frictional characteristics of the material. The investigation proved the following points:

1. The assumption about the possibility of omitting the final reduction of metal powders was found correct.
2. The use of as-delivered powders simplify the technology of manufacture (the operations of final reduction and comminution fall away) and lowers the production costs.
3. The quality of sintered friction parts does not deteriorate when as-delivered powders are used.

The Effect of Variables in the Operations of Pressing and Sintering on the Physicomechanical and Frictional Properties of FMK-11

Variables in Pressing Operation

The selection of a rational manufacturing process must be made with the aim of achieving the best mechanical and frictional properties of the final product. A lack of sufficiently reliable data on the effect of compacting pressure on the frictional properties has made the following investigation necessary. Physical, mechanical, and frictional properties of specimens compacted at speci-

Table 2

Content of oxygen in the parent iron powder, wt.%	Coefficient of friction at									
	100°C	200°C	300°C	400°C	500°C	600°C	700°C	800°C	900°C	1000°C
0.7	0.63	0.57	0.49	0.43	0.37	0.32	0.26	0.23	0.20	0.20
1.5	0.63	0.56	0.48	0.45	0.39	0.36	0.29	0.26	0.22	0.22
2.0	0.63	0.56	0.49	0.45	0.44	0.39	0.29	0.25	0.21	0.19
3.5	0.63	0.58	0.52	0.45	0.41	0.34	0.28	0.25	0.23	0.21
5.4	0.61	0.56	0.49	0.46	0.44	0.36	0.32	0.26	0.24	0.20

Table 3

Specific compacting pressure, Mp/cm^2	Brinell hardness	Wear, mm	
		FMK-11	Cast iron
2	86	0.39	0.46
4	90	0.55	0.18
6	93	0.25	0.21
8	94	0.32	0.21

fic loads of 2, 4, 6, and 8 Mp/cm^2 were determined. Brinell hardness and linear wear values for one complete test cycle are given in Table 3. Hardness was measured on 24 specimens, frictional properties on 6 specimens.

The following conclusions can be drawn:

1. Hardness increases as compacting pressure increases.
2. An increase in the specific compacting pressure from 2 to 6 Mp/cm^2 lowers the wear of the sintered specimens by a factor of 1.5 and the wear of the cast iron by a factor of 2. Further increase to 8 Mp/cm^2 causes an insignificant rise in the wear of the sintered material, presumably due to overpressing and separation of layers. The wear of cast iron remains unchanged.

The coefficient of friction changes but little with respect to different compacting pressures (see Table 4). However, in the temperature range from 600 to 1000°C, specimens compacted at a specific pressure of 6 Mp/cm^2 exhibit the highest values of the friction coefficient.

Table 4

Compacting pressure, Mp/cm^2	Coefficient of friction at									
	100°C	200°C	300°C	400°C	500°C	600°C	700°C	800°C	900°C	1000°C
2	0.65	0.58	0.51	0.45	0.38	0.34	0.30	0.26	0.23	–
4	0.61	0.56	0.52	0.48	0.47	0.40	0.31	0.24	0.24	0.20
6	0.58	0.49	0.49	0.52	0.45	0.44	0.35	0.26	0.26	0.24
8	0.63	0.60	0.54	0.53	0.46	0.40	0.34	0.25	0.23	0.20

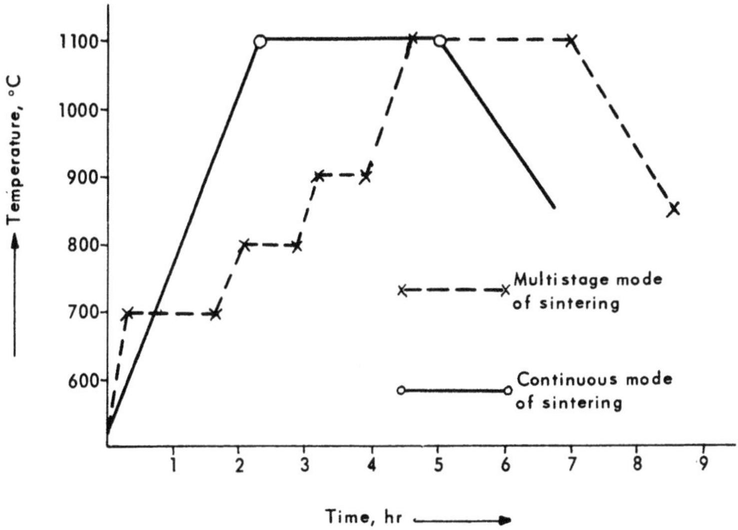

Fig. 2

The results point to the following conclusions:

1. The increase in specific compacting pressure up to 6 Mp/cm^2 raises the density and hardness of the sintered friction material.
2. The material pressed at 6 Mp/cm^2, exhibits the lowest wear.

Variables in Sintering Operation

The following multistage mode of heating was formerly used in the sintering operation of the friction material (Fig. 2):

1. Charging the material into the furnace at 600°C.
2. Heating to 700°C, keeping the temperature constant for 1.5 hours.
3. Heating to 800°C, keeping the temperature constant for 45 minutes.
4. Heating to 900°C, keeping the temperature constant for 45 minutes.
5. Heating to 1040-1060°C, keeping the temperature constant for 2.5 hours.
6. Cooling in furnace to 800-850°C.

This multistage sintering system requires a programmed operation of the furnace and prolongs the production cycle (the total time of sintering is 7 hours). It was used to ensure a uniform distribution of temperature within all sintered parts and to lower the decarburization of the steel backing.

The presence of an underlayer in the products made from the FMK-11 sintered material protects the steel backing against decarburization. It was suggested to simplify the sintering system by adopting a continuous temperature rise up to the desired temperature of sintering. The new system is also shown in Fig. 2. It is readily seen that the total sintering time has dropped to 4.5–5 hours.

The adoption of the simplified sintering system does not lower the quality of the material, and no carburizing or embrittlement of the steel backing has been observed. It has shortened the actual time of heating to the sintering temperature from 4.5 to 2-2.5 hours and has simplified the operation. This process has been used in the production of the sintered friction materials.

Application of Load during Sintering

Sintering takes place under an applied load in order to attain high hardness, required frictional properties, and a reliable adhesion of the sintered layer to the steel backing. In order to evaluate the effect of applied load, the density, hardness, and frictional properties of specimens sintered under specific loads of 15, 20, 25, and 30 kp/cm^2 were measured. The specific compacting pressure was 6 Mp/cm^2 in all cases.

Table 5

Specific load applied during sintering, kp/cm^2	Density, g/cm^3	Hardness, kp/mm^2	Wear, mm	
			FMK-11	Cast iron
15	5.6	90	0.44	0.20
20	5.7	92	0.30	0.20
25	5.9	97	0.30	0.21
30	5.9	97	0.34	0.32

Table 6

Specific load applied during sintering, kp/cm²	Coefficient of friction at									
	100°C	200°C	300°C	400°C	500°C	600°C	700°C	800°C	900°C	1000°C
15	0.62	0.58	0.53	0.49	0.45	0.45	0.29	0.24	0.21	0.20
20	0.61	0.56	0.49	0.46	0.40	0.32	0.28	0.25	0.22	0.19
25	0.63	0.58	0.52	0.52	0.39	0.35	0.31	0.28	0.21	0.19
30	0.64	0.59	0.53	0.53	0.41	0.34	0.28	0.23	0.21	0.19

The values of density, hardness and linear wear as related to the applied sintering load are given in Table 5. The values are the average results of six measurements. The data indicate that density and hardness increase with increasing sintering load up to 25 kp/cm^2. The wear of the sintered material decreases roughly 1.5 times; the wear of the cast iron remains unchanged. Further rise in sintering load up to 30 kp/cm^2 does not increase the density and hardness any more, but wear of the FMK-11 increases a little and that of the cast iron increases by 60%. At the same time, the application of the highest specific load causes a significant deformation of specimens accompanied by appearance of cracks. The coefficient of friction remains unchanged, apparently independent of the investigated variable factor (Table 6). The data are again the average results of six experiments. The adhesion of the sintered layer to the steel backing was checked by bending at the angle 45° between the sintered layer and a horizontal surface. The quality of adhesion improved with increase in specific sintering load.

It can be concluded that specific loads from 20 to 25 kp/cm^2 applied during the sintering process give the best results and, therefore, are at present being used in current production.

Selection of a Suitable Paint for Interlaying Discs, Preventing Brazing during Sintering

Beside research programs directed toward improvements of the quality of production, tests have been performed with the aim of improving the working conditions and raising the productivity of labor. One of these tasks is the selection of a suitable antibrazing paint.

According to the present technology the compacts are sintered in packets. A packet consists of compacts successively interlayed by steel discs. Until our investigation was performed, paper and loose graphite had been inserted between discs.and compacts to prevent brazing to one another. The disadvantages of this method are obvious: contamination of the equipment by graphite powder and impossibility of mechanizing the process. Therefore an effort was made to replace the paper and graphite by a suitable paint.

An antibrazing paint must meet the following requirements:

1. Adhere properly to the disc and readily form a uniform layer.
2. Dry quickly at room temperature.
3. Not react with any component of the compacts.
4. Protect the compacts against brazing to the steel discs.

The following compositions of paint were tested:

1.	Zircon concentrate, iron-free.	37%
	Silica flour	30%
	Soluble glass (density 1.12-1.13)	33%
2.	Hydrolyzed solution of ethyl silicate	1 part
	Calcinated silica flour	2 parts
3.	Water suspension of lime	
4.	Water suspension of chalk	
5.	Carbon black.	22-24%
	Silica sand	14-16%
	Refractory clay.	9-10%
	Balance water	

Paints Nos. 1 and 2 were put on a sand-blasted surface of the steel discs. Paints Nos. 3, 4, and 5 were applied to the discs without any preceding surface preparation. The drying took place in air.

The testing of the described paints has given these results:

1. Paints Nos. 3 and 4 do not prevent brazing between the compacts and discs.

2. Paints Nos. 1 and 2 prevent brazing but have several significant disadvantages:
   a. Paint No. 1 sticks to the surface of the compacts after sintering. A continuous layer of the paint is formed on the surface which is difficult to remove.
   b. Paint No. 2 adheres poorly to the surface of the discs and peels off on drying. Application of the paint to preheated discs improves adhesion a little, but does not prevent peeling completely.
3. The best results were obtained using paint No. 5, which best meets the above requirements.

Paint No. 5 can be fully recommended for the purpose of covering the interlaying discs. It does not require any special surface treatment of the discs and reliably protects them from brazing. Its use makes it possible to mechanize the operation of painting and to improve industrial hygiene. The paint is currently being tested under actual production conditions.

Chapter 11

The Influence of Copper on the Properties of Sintered Iron-Graphite Friction Materials

V. Procházka, E. Navara, V. Miskovic, and K. Kosta

Laboratory of Metallurgical Technology
Technical University
Košice, Czechoslovakia

Introduction

The high requirements engineering places on friction elements has lead to the development and production of friction materials based on metals. Research in recent years has paid much attention to iron—graphite friction materials. For example, Eisenkolb [1], Strobel [2], Dufek [3], and others have dealt with these problems extensively.

An important component of iron-based sintered friction materials are nonferrous metals, such as copper or copper alloys. Extensive basic research has been going on at our laboratory concerning steel sintered friction materials.

After examining the effect of graphite [4], baryte [5], and aluminum oxide and silicon oxide [6], the effect of copper was also investigated. The effect of copper powder on the properties of P/M steel is the subject of several research publications, of which Goncharov [7] and Aksenov and Sorokin [8] will be cited. Problems concerning the effect of copper, mainly on sintered antifriction materials, have been tackled by Domsa and his colleagues [9].

The effect of the addition of copper powder on the friction properties of P/M steel has also been studied by Bebnev [10]. The present literature does not give a complete picture of the effect of copper on friction materials. Copper affects the heat conductivity of the material, has a graphitizing effect on the Fe–C alloys, affects the dissolving of carbon in iron, the kinetics of the sintering of the Fe–Cu–C system, and, in the presence of additives, can also react with other components. The addition of copper can have both positive and negative effects on the friction properties of P/M steel. Therefore it is necessary to evaluate the addition of copper. This paper is meant to be a contribution to these problems and is aimed at the investigation of the addition of up to 15% copper powder.

Experiments

For the purpose of investigating the effect of different additions of copper, samples containing 1.5%, 3%, 5%, 10%, and 15% Cu

Table 1. Marking and Composition of Samples Containing Different Amounts of Copper

Sample No.	Added amounts, %					
	Graphite	$BaSO_4$	SiO_2	Al_2O_3	Cu	Fe
1	6	14	4	–	1.5	Balance
2	6	14	4	–	3.0	"
3	6	14	4	–	5.0	"
4	6	14	4	–	10.0	"
5	6	14	4	–	15.0	"
6	6	14	–	5	1.5	"
7	6	14	–	5	5.0	"
8	6	14	–	5	5.0	"
9	6	14	–	5	10.0	"
10	6	14	–	5	15.0	"

were produced. According to the results of earlier research [6], two basic materials were chosen: one containing aluminum oxide, the other containing silicon oxide. The mixture containing silicon oxide, which will be designated as A, contained 6% graphite, 14% $BaSO_4$, 4% SiO_2, and the balance powdered steel. The mixture containing aluminum oxide, designated as B, contained 6% graphite, 14% $BaSO_4$, 5% Al_2O_3, and the balance powdered steel. The contents of the different samples are shown in Table 1. The powdered materials were mixed in drums for 4 hours. From this mixture cylindrical specimens 30 mm in diameter and 10 mm high were compacted. These cylinders were then sintered in a silite retort furnace in the following manner.

Up to a temperature of 700°C, dried hydrogen was brought into the retort. Above this temperature, nitrogen was used. The sintering temperature of 1000°C was maintained for 2 hours. After cooling to 700°C, nitrogen was again replaced by hydrogen as a protection against oxidation by oxygen contained in the nitrogen. This process will subsequently be referred to as sintering in a combined atmosphere.

The sintered specimens were then subjected to mechanical tests and measurement of friction properties on SFM-2 equipment. This device was made at the Laboratory of Metallurgical Technology in Košice and is a modified version of the well-known Soviet instrument I-47. The SFM-2 device is shown and described in another paper [11].

During the friction tests, the sample is first run-in at low rpm until the friction coefficient stops fluctuating rhythmically. The actual test is done at 100-5000 rpm with the following circumferential velocities: 0.13, 0.26, 0.63, 0.91, 1.3, 1.88, 2.6, 3.25, 3.9, 5.2, and 6.5m/sec.

Each sliding speed is always maintained for 15 min. The temperature and friction coefficient are measured every 5 minutes. Temperatures are measured by a thermocouple fitted into a 1.2-mm-diameter hole in the sample. The joint of the thermocouple is 1 mm from the friction surface. After the test, the sample is measured by weighing and measuring to determine the loss of material or the linear and weight wear.

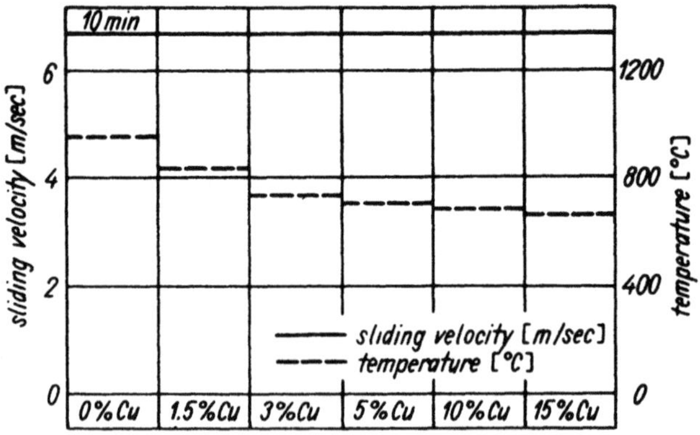

Fig. 1. Effect of the addition of Cu on the thermal durability of samples made from mixture A.

Tests of the friction properties of samples Nos. 1, 2, 3, 4, and 5, which were made from mixture A, show that thermal durability after an addition of copper powder does not change and all samples withstand the whole test as can be seen from Fig. 1. When the highest sliding speed of 6.5 m/sec is used, the measured temperature decreases with the addition of copper from 810°C to 660°C. This is probably due to an improvement of thermal conductivity of the sample with the rising content of copper. It can be confirmed by measuring the specific resistance, shown in Table 2, which de-

Table 2. Specific Electrical Resistance of Samples Containing Different Amounts of Copper

Specimens from mixture A		Specimens from mixture B	
Addition of copper, %	Specific electric resistivity, $\Omega \cdot mm^2/m$	Addition of copper, %	Specific electric resistivity, $\Omega \cdot mm^2/m$
0	0.4031	0	0.3638
1.5	0.3643	1.5	0.3257
3.0	0.3522	3.0	0.3164
5.0	0.3405	5.0	0.3097
10.0	0.3132	10.0	0.2942
15.0	0.2913	15.0	0.2732

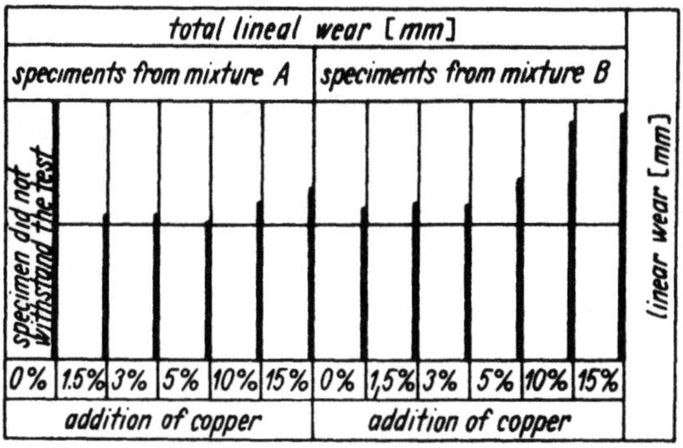

Fig. 2. Total linear wear according to additions of powdered copper.

creases from 0.3643 $\Omega \cdot mm^2/m$ with samples containing 1.5% Cu to 0.2913 $\Omega \cdot mm^2/m$ with samples containing 15% Cu.

The total linear wear practically does not change with the content of Cu, and stays between 1.02 mm and 1.27 mm.

Samples with a content of 10 and 15% of copper show higher thermal wear than samples with a lower copper content. Total linear wear of samples with different Cu content is shown in Fig. 2.

The friction coefficient gradually decreases with higher amounts of copper. For example, at a sliding speed of 0.26 m/sec and a 1.5% content of Cu, its value is 0.51; this drops to 0.42 when the content of Cu reaches 15% (see Fig. 3). At higher speeds, say,

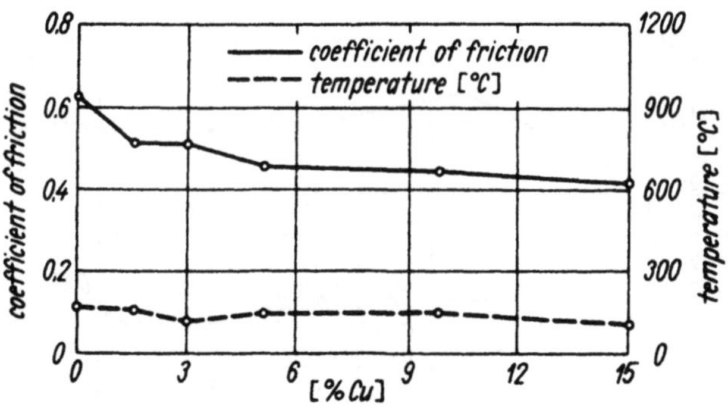

Fig. 3. Dependence of the friction coefficient and the temperature on the percentage of Cu at a 0.26 m/sec sliding speed. Composition: 6% graphite, 14% $BaSO_4$, 4% SiO_2, balance powdered steel.

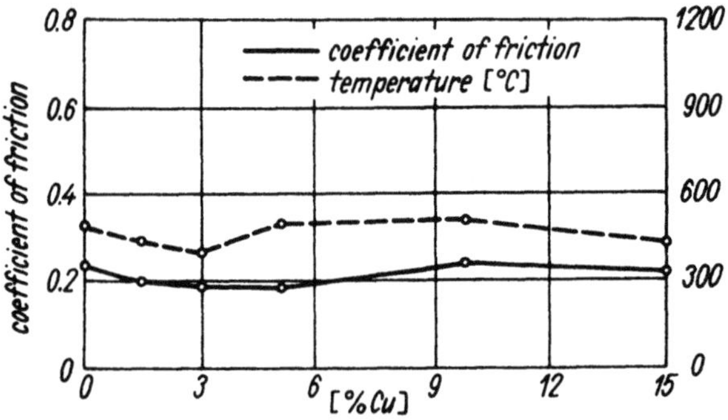

Fig. 4. Dependence of the friction coefficient and the temperature on the percentage of Cu at a 2.6 m/sec speed. Composition: 6% graphite, 14% $BaSO_4$, 4% SiO_2, balance powdered steel.

2.6 m/sec, the friction coefficient practically does not change and remains between 0.23-0.19 (see Fig. 4). The friction temperature at a sliding speed of 0.26 m/sec stays within the limits of 120°C and 160°C.

Compressive strength first increases with the addition of copper from 31.0 kp/mm² with 1.5% Cu to 37 kp/mm² with 5% Cu. With higher percentages of Cu, the compressive strength settles at 33.5 kp/mm². Brinell hardness shows a similar curve when reaching a value of 66 kp/mm² with samples containing 5% Cu and slightly dropping to 58.2 kp/mm² at 15% Cu. A mutual dependence of mechanical properties and specific electrical resistance is shown in Fig. 5.

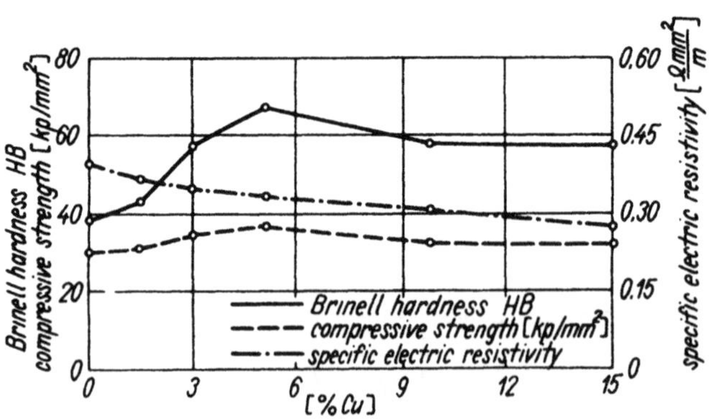

Fig. 5. Effect of the addition of Cu on the hardness, pressure strength, and specific electrical resistance. Composition: 6% graphite, 14% $BaSO_4$, 4% SiO_2, balance powdered steel.

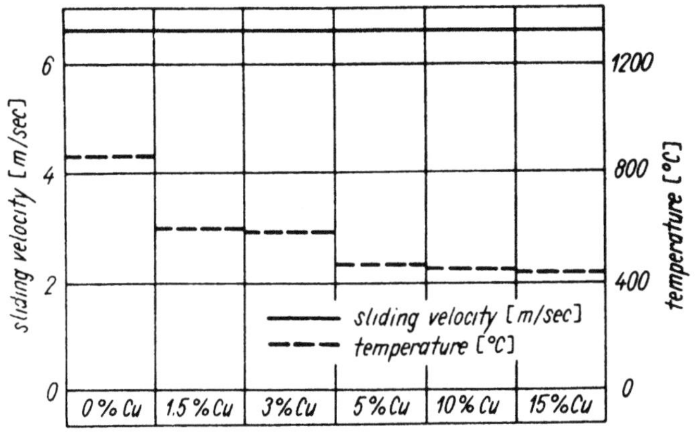

Fig. 6. Effect of the addition of Cu on the heat conductivity of samples made from mixture B.

Samples Nos. 6, 7, 8, 9, and 10, which were made from mixture B containing 6% graphite, 14% $BaSO_4$, and 5% Al_2O_3, also showed thermal stability which did not change with the addition of Cu. All samples withstood the complete test on the SFM-2 equipment, as can be seen from Fig. 6. The friction temperature gradually decreased from 600°C with 1.5% Cu down to 450°C with 15% Cu, which was due to an improvement in the heat conductivity of the samples.

An improvement of heat conductivity is proved by measurements of electrical resistance, which are shown in Table 2. Speci-

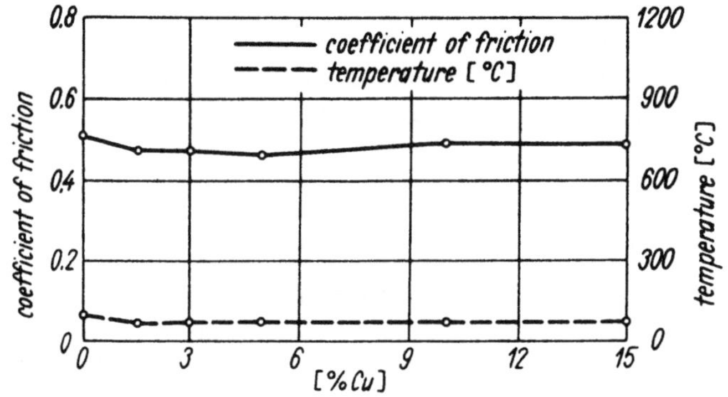

Fig. 7. Effect of the addition of Cu on the friction coefficient and the temperature at a 0.26 m/sec speed. Composition: 6% graphite, 14% $BaSO_4$, 5% Al_2O_3, balance powdered steel.

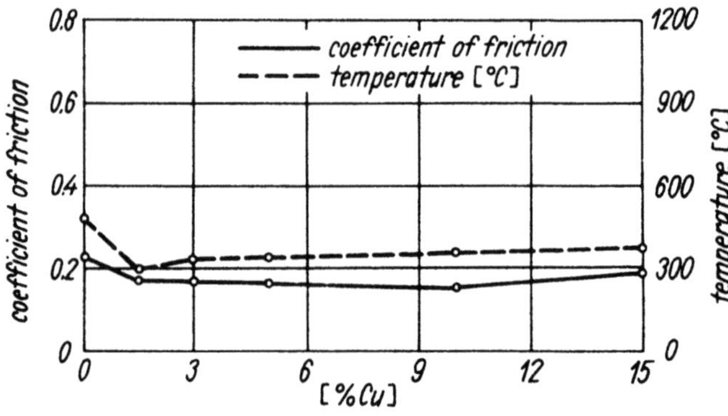

Fig. 8. Effect of the addition of Cu on the friction coefficient and the temperature at a 2.6 m/sec speed. Composition: 6% graphite, 14% $BaSO_4$, 5% Al_2O_3, balance powdered steel.

fic electrical resistance dropped from a value of 0.3257 $\Omega \cdot mm^2/m$ with 1.5% Cu to 0.2732 $\Omega \cdot mm^2/m$ with 15% Cu.

Total linear wear increases with higher percentages of Cu from 1.22 mm at 1.5% Cu to 1.88 mm at 15% Cu. The effect of different percentages of copper on total linear wear is shown clearly in Fig. 2.

The coefficient of friction at a sliding speed of 0.26 m/sec practically does not change with the addition of copper, as can be seen from Fig. 7. With samples containing 1.5% copper, its value is 0.48, and with 15% copper 0.50. The behavior of friction temperatures, which stay between 110°C and 140°C , is similar. When the sliding speed is increased to 2.6 m/sec, the addition of copper to the basic material has again no practical effect on the coefficient of friction, which, with 1.5% content of copper, has a value of 0.18, and with 15% content of copper a value of 0.19. These relations are shown in Fig. 8.

The friction temperature has a slightly increasing tendency. Sample No. 6 reached 292°C while sample No. 10 reached 360°C.

Compressive strength slightly increases from 51 kp/mm^2 with sample No. 6 to 52.8 kp/mm^2 with sample No. 7. Further additions of copper result in a small drop of compressive strength. Sample No. 10, which is a 15% content of copper, has only 41 kp/mm^2. Hardness varies in a similar manner and reaches a maximum of

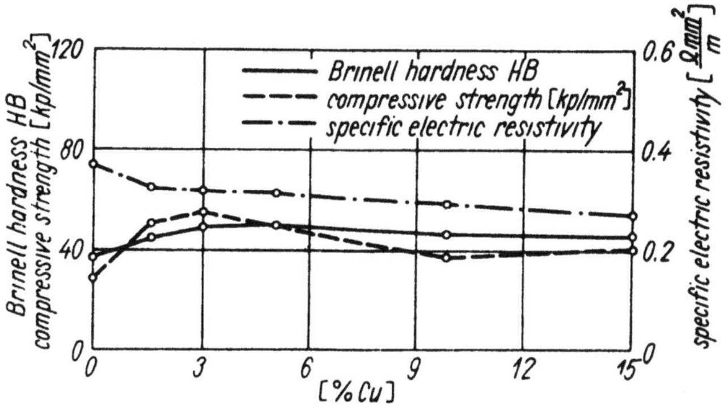

Fig. 9. Effect of the addition of Cu on the hardness, pressure strength, and specific electrical resistance. Composition: 6% graphite, 14% $BaSO_4$, 5% Al_2O_3, balance powdered steel.

49 kp/mm² with sample No. 7 and shows a dropping tendency with further additions of copper. Sample No. 10 displays a hardness of 46.8 kp/mm². The dependence of hardness and pressure strength on the content of copper is shown in Fig. 9.

To determine the effect of copper on changes in multicomponent specimens during sintering, chemical analyses were performed to investigate carbidic carbon in all examined materials. The results are shown in Tables 3 and 4.

Table 3. Analyses of Chemically Bonded Carbon in Samples Made from Mixture A

Addition of copper, %	Carbon, chemically bonded, %
0	1.36
1.5	1.32
3.0	1.35
5.0	1.36
10.0	1.28
15.0	1.22

Table 4. Analyses of Chemically Bonded Carbon in Samples Made from Mixture B

Addition of copper, %	Carbon, chemically bonded, %
0	1.41
1.5	1.20
3.0	1.22
5.0	1.18
10.0	1.14
15.0	1.13

Analyses have proved that with materials containing SiO_2, copper does not affect the amount of carbidic carbon up to a content of 5%. Higher additions of copper usually lower the amount of carbidic carbon, as shown in Table 3. Copper added to materials containing Al_2O_3 definitely lowers the amount of carbidic carbon, as shown in Table 4.

Evaluation of Experimental Results

The complete set of experimental results shows, as had been expected, that copper has both good and negative effects. The reasons can be found in the fact that copper during sintering dissolves in iron in accordance with the binary Fe–Cu diagram. Copper thus affects the relations between iron, bound carbon, and possibly also dissolved silicon.

According to the binary diagram, α-iron dissolves up to 2% Cu at a temperature of 850°C, which is at the same time the temperature of the eutectoidal change from α to γ. γ-iron at this temperature dissolves 4% of copper (eutectoidal composition). By increasing the temperature, the solubility of copper in austenite reaches a value of 8% at 1100°C.

The solubility of carbon in copper-saturated austenite, that is, in a ternary solid solution, is much lower than in ordinary austenite [9].

A lower content of dissolved carbon obstructs or completely stops the forming of secondary cementite and lowers the amount of bound carbon in the final structure of the material. This is suitable for self-lubricating bearings or other parts which are subject to friction at lower temperatures, but it lowers the resistance of material to thermal wear.

The above analysis of the process of sintering the composition Fe–Cu–C is applicable to material containing Al_2O_3. As is shown in Table 4, chemical and metallographic tests established a lower amount of bound carbon, which is probably the reason for a lower resistance against wear at high temperatures, especially with higher contents of copper.

The presence of a free copper phase in the structure, which, for 5% or more Cu, is observable with the aid of a microscope by its red coloring, leads to a deterioration of frictional properties. This phase in itself is not resistant against wear at high temperatures.

The addition of copper to material made from mixture B brings one advantage – the improvement of heat conductivity – and two disadvantages – a lowered solubility of carbon in austenite and, at higher temperatures, a lower resistance to thermal wear. Practical results of friction tests point out that the disadvantages seem to outweigh the advantage, especially when the addition of copper reaches 5% and above. Only additions of copper up to 3% seem to be practicable for improving the hear conductivity of the material.

A more complicated situation seems to appear with materials containing SiO_2. In an earlier report [6], it has been proved that silicon oxide is partly reduced by carbon during sintering while silicon dissolves in iron.

Dissolved silicon closes the γ region and lowers the solubility of carbon limited by curve SE in the iron–carbon diagram [12]. For this reason the level of bound carbon drops with the addition of SiO_2, as shown in Table 3.

A comparison of the analysis of bound carbon in Tables 3 and 4 shows a remarkable fact, that the level of carbidic carbon in the presence of copper and silicon is higher than in the presence of copper only, even though silicon displaces carbon even more radically than copper [3]. Besides that, the level of carbidic carbon practically does not drop with the addition of copper up to 5% in materials made from mixture A. Only maximum additions of copper bring about a slight decrease.

This can be explained by the fact that the effect of copper is considerably reduced by the presence of silicon. This can be supported by the binary Cu–Si diagram, which shows a high solubility of silicon in copper (5.4% at 830°C) and also a low melting temperature of the Cu–Si alloy. It can be expected then, even though no results can be found in the published literature, that the diffusion of silicon in copper is easier than in iron under the sintering conditions of our materials. Also the results of measuring the electrical conductivity and their comparison with results of the same measurements made on material B with the same percentage of copper show, as can be seen in Table 2, that copper in mixture A has a lower heat and electrical conductivity, which indicates that it contains a dissolved component, which can only be silicon (a certain small amount of iron has to be taken into consideration in both cases). The dissolving of silicon in copper may then influence not only the dissolving of silicon in iron, but also slow down the process of the diffusion of copper into iron. Practical tests show that a 5% addition of copper gives optimum results.

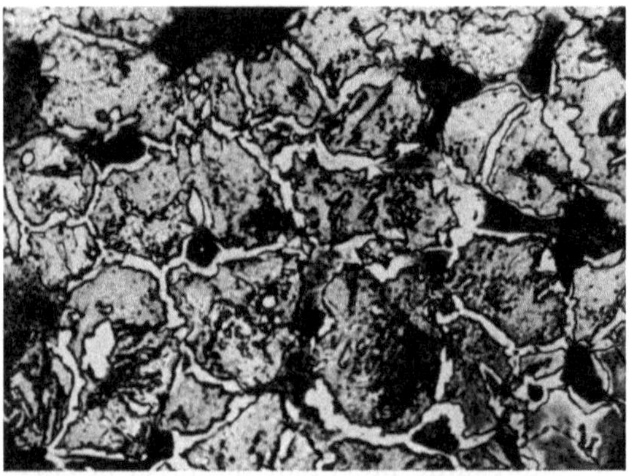

Fig. 10. Micrograph of a sample containing 6% graphite, 14% $BaSO_4$, 4% SiO_2, 3% Cu, balance powdered steel, etched with nital, magnified 500 ×.

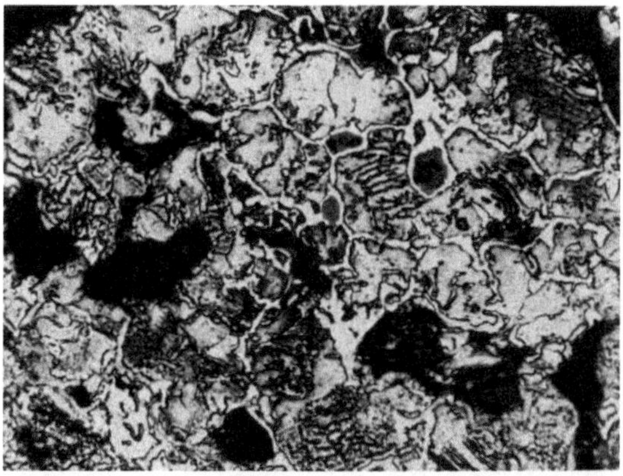

Fig. 11. Micrograph of a sample containing 6% graphite, 14% $BaSO_4$, 10% Cu, balance powdered steel, etched with nital, magnified 500 ×.

A detailed metallographic analysis confirmed all conclusions on the influence of copper on both types of sintered materials. A test of microhardness made on many copper-phase grains showed a relative higher hardness of the grains of material from mixture. A. (The ratio of diagonals of hardness indentations made with a 2 g force was 12 : 16).

The effect of copper on the amount of cementite is illustrated in Figs. 10, 11, 12, and 13. The first two show a very similar structure, even though the addition of copper is 3% in the first and 10% in the second case. In both cases samples are from mixture A. The second pair of samples (made from mixture B) shows a considerable decrease of cementite with increase of copper from 3% to 10%. Figure 13 shows a complete breakdown of the cementite network with only isolated groups of cementite grains observed.

Conclusion

It has been proved that the thermal durability of material containing silicon oxide increases with the addition of copper. In this case the positive influence of copper prevails. An increase of thermal wear is noticeable only with higher percentages of copper and therefore, in accordance with metallographic and

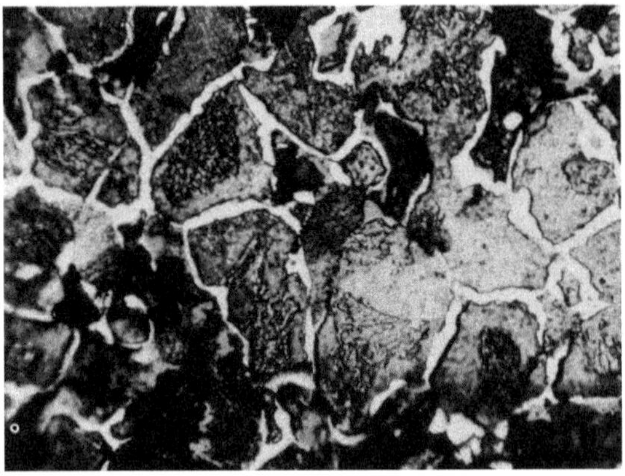

Fig. 12. Micrograph of a sample containing 6% graphite, 14% $BaSO_4$, 5% Al_2O_3, 3% Cu, balance powdered steel, etched with nital, magnified 500×.

chemical analyses, it can be said that the optimum content of copper in materials containing SiO_2 is roughly 5%.

Additions of copper to materials containing aluminum oxide do not increase thermal durability. Percentages of copper higher than 3% leads to an increase in thermal wear.

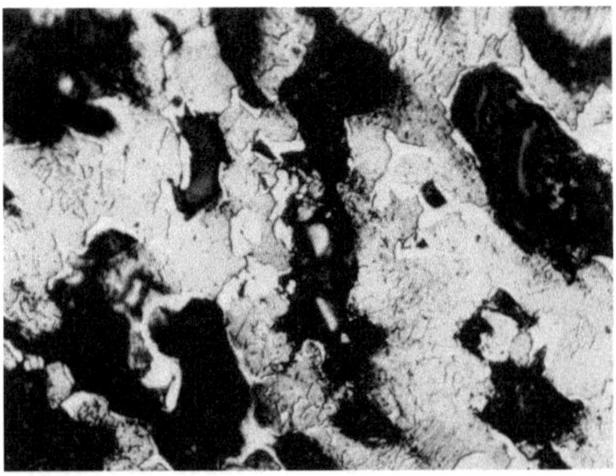

Fig. 13. Micrograph of a sample containing 6% graphite, 14% $BaSO_4$, 5% Al_2O_3, 10% Cu, balance powdered steel, etched with nital, magnified 500×.

References

1. Eisenkolb, F., Über pulvermetallurgisch hergestellte Gleit-und Reibwerkstoffe, Akademie-Verlag, Berlin, 1958.
2. Strobel, E., and Übel, H., Proceedings of the 2nd International Powder Metallurgy Conference in Eisenach, Akademie-Verlag 1962, pp. 541-550.
3. Dufek, V., Kovokeramické třecí materiály. SNTL, Prague, 1964.
4. Procházka, V., and Miškovič V., Kovové materiály, No. 2, pp. 408-418, 1964.
5. Procházka, V., Navara, E., and Miškovič, V., Pokroky práškové metalurgie, No. 3, pp. 28-41, 1965.
6. Procházka, V., and Navara, E., this volume, Chapter 6.
7. Goncharova, V. M., Powder Metallurgy, Yaroslavl, p. 61, 1956.
8. Aksenov, G. I., and Sokolin, V. K., Powder Metallurgy, Yaroslavl, p. 71, 1956.
9. Domsa, A., Neue Hütte No. 9, pp. 186-188, 1964.
10. Bebnev, P. I., Powder Metallurgy, Yaroslavl, p. 233, 1956.
11. Procházka, V., and Miškovič, V., Pokroky práškové metalurgie, No. 3, pp. 3-14, 1964.
12. Houdremont, E., Hanbuch der Sonderstahlkunde, Part II, Springer Verlag, 1956.
13. Kontorovič, I. E., Tepelné spracovanie ocele a liatiny, P. V., Prague, 1952.

Chapter 12

New Sintered Friction Materials

I. M. Fedorchenko and V. M. Kryachek

Institute of Metal
Ceramics and Special Alloys
Academy of Sciences of the Ukrainian SSR
Kiev, USSR

Contemporary engineering makes extensive use of sintered friction materials; they are used in a variety of applications, either for transmission of motion through friction or for deceleration and braking. These materials are based on iron or copper; a typical composition of copper-based friction materials lies within the following percentage ranges of the components: tin, 8-10%; lead, 5-12%; iron, 2-8%; graphite, up to 8%; zinc, up to 7%; asbestos, silica, metallic oxides, possibly sulfides and/or carbides, up to 10%; balance copper.

Copper-based materials are used mainly where boundary or semifluid friction occurs. For dry friction they are suitable only in cases where operation conditions are relatively mild, especially from the viewpoint of thermal regime.

The properties, applications, fabrication, and advantages of copper-based sintered friction materials have been extensively dealt with in the literature, (see [1-4]). Copper-based materials, beside a number of advantages, possess several disadvantages:

1. Insufficient coefficient of friction in the presence of a lubricant at high sliding speeds, which leads to undesirably large sizes of frictional elements.
2. Insufficient wear resistance at high loads and sliding speeds.
3. Limited possibilities of application in applications where dry friction occurs (medium loads, speeds, temperatures).
4. The materials contain expensive and deficient tin, the cost of which is up to 50% of the total price of the product (if the content of tin does not exceed 10 wt.%).

A new copper-based sintered material, which has been developed at the Institute of Metal Ceramics and Special Alloys of the Ukrainian Academy of Sciences, is being tested under actual operation conditions. Its advantages are that it does not contain tin, which is expensive, and that it has properties that are superior to those of existing friction materials.

The fabrication of this new material is not substantially different from that of conventional friction materials. It only requires a somewhat higher sintering temperature (950–980°C) and pressure during sintering (20–30 kp/cm^2). Characteristics of the material are given in Table 1.

Frictional properties were investigated both under conditions of dry friction and with the use of a lubricant. The coefficient of friction and wear resistance were controlled by varying the contents of nonmetallic hard components, so-called frictional additives.

Table 2 gives the results of an investigation of several alloys with different contents of frictional additives. Tests were performed with 40 × 45 × 5 mm specimens (3-mm steel backing and 2-mm frictional layer) manufactured by pressure sintering as described above. The specimens were then polished and grooves were milled in the surface of the sintered layer to allow the flow

Table 1

Property of the frictional material	Existing copper-based frictional materials	New friction materials
Brinell hardness, kp/mm^2	25-50	60-100
Shearing strength, kp/mm^2	8-10	15-20
Bending strength, kp/mm^2	10-15	20-30
Cost, rubles pkg	2.0-2.3	1.0-1.1

Table 2

Alloy No.	Brinell hardness, kp/mm^2	Coefficient of friction at the velocity (m/sec)						Linear wear after one cycle, μ
		40	30	20	10	5	2	
1	65-70	0.079	0.077	0.078	0.082	0.079	0.091	4-8
2	73	0.092	0.081	0.074	0.086	0.082	0.074	17
3	83	0.083	0.080	0.085	0.098	0.098	0.106	21
4	54-63	0.100	0.087	0.082	0.081	0.080	0.087	4-10
5	67-80	0.090	0.092	0.085	0.086	0.084	0.094	6-13
6	74-90	0.095	0.093	0.091	0.093	0.089	0.099	8-18
Alloy MK-5	35-50	0.084	0.081	0.071	0.068	0.066	0.070	10-20

of oil and the removal of wear products. The grooves were 1 mm wide, 0.8 mm deep, and 2 mm apart. Their slope was 5° toward the direction of friction. Before the test the specimen was pressed and conformed to the disc (made from grade 60G steel). During the test, the surface of the disc was lubricated (MT-16P oil). All tests were performed under a specific load of 30 kp/cm^2 and at a constant sliding velocity. The range of applied velocities was from 2 to 40 m/sec.

Three tests were made at each velocity, two lasting 30 seconds, which corresponds to a short-time process of braking or clutching (nonstationary thermal regime), and one lasting 120 sec, which measures the coefficient of friction under stabilized thermal conditions. Between tests, the specimen was cooled to an initial temperature of 70-75°C. The temperature was measured by a thermocouple at the center of the specimen at a distance of 1 mm from the rubbing surface. During the tests, the torque, specific load, temperature of the specimens, and sliding velocity were measured.

Specimen wear was calculated from the measurement of thickness at two fixed points before and after the testing at the entire range of velocities. The measurements were performed with an accuracy of $\pm 2\ \mu$.

The values of wear given in the Table 2 are average values, relating to one cycle of the full range of velocities. The state of the rubbing surface of the specimen and the steel disc was controlled visually. The new alloys exhibit a good conformability; their surface is smooth without scratches and pluck-off spots after the experiments.

The surface of the steel disc is covered with a smooth film without grooves and traces of seizure. The specimens made from the new alloys usually performed noiselessly and without vibrations.

In Table 2, the properties of the common frictional alloy MK-5 are given, for the sake of comparison. The coefficient of friction of the new alloys is a little higher than that of MK-5. They also exceed the MK-5 alloy in wear resistance by a factor of two.

The new materials are now being tested in the brakes and clutches of tractors and other machines. Some of them have passed laboratory and operational tests in friction elements operating without lubrication. They have also operated satisfactorily under conditions of rolling friction (gear boxes) and sliding friction (clutches and brakes). These new friction materials can increase the service life and reliability of friction elements, lower their cost, and simplify their maintenance. For example, the application of the friction material for lining the surface of cylinders of a friction drive (variable-speed gear box) after Svetovarov, has made it possible to lower the cost of the friction aggregate, to increase the transfer of power from 3 to 10 kW, to raise the transmitted number of revolutions per minute, to prolong the life from 300-500 hr to 7000-8000 hr, and to cut the maintenance cost substantially [5]. The new friction material is also used in the disc brakes of vulcanizing machines produced in a plant in Kiev.

References

1. Balshin, M. J., Powder Matallurgy, Mashgiz, 1948.
2. Panaioti, I. I., Fedorchenko, I. M., and Skorokchod, V. V., Mashinostroenie i priborostroenie, No. 10-11, Kiev, 1958.
3. Properties and Technology of Manufacturing Metalloceramic Products, Kiev-Leningrad, 1960.
4. Lvovskij, M. M., and Goreyev, R. N., Mashinostroenie, No. 2, Kiev, 1963.
5. Kryachek, V. M., and Kosterov, P. M., Mashinostroenie, No. 2, Kiev, 1963.

Chapter 13

Studies of Friction Materials

I. M. Fedorchenko, I. I. Panaioti, G. M. Derkacheva,
I. Ya. Dzykovich, and G. N. Gordan

Institute of Materials Science
Academy of Sciences of the Ukrainian SSR
Kiev, USSR

Part I

Asbestos-rubber and asbestos-resin friction compositions are currently used in the friction units of automobiles, tractors excavators, and other machines. The low wear resistance of these materials, due to a deterioration of the physicomechanical properties of the binder (rubber, resin) when the friction surface temperature rises during the operation of the friction pair, is their major drawback. In view of this, such friction materials cannot secure the required service characteristics in the friction units of modern machines and mechanisms.

Frequently, sintered copper-base friction materials give a satisfactory performance. The main disadvantages of these materials are their unsuitability for operation at temperatures above 350°C, high wear in operation under dry friction conditions, as well as the fact that their components are in short supply.

There is thus an urgent need for the development of friction materials based on a more readily available metal, such as iron, which would satisfy the requirements of the friction units of modern machines. At the present time, FMK-11 is one of the best sintered iron-base friction materials made in the USSR. However, although its friction properties are superior to those of asbestos-rubber and asbestos-resin materials, it is not sufficiently wear resistant, and for this reason is frequently not suitable for heavy-duty operation. In recent years, the foreign literature, mainly patents, has presented some very scanty information, frequently without indication of compositions or manufacturing procedures, on researches being carried out with the aim of developing sintered iron-base friction materials [1-4]. The latter are generally iron-graphite compositions containing up to 50 vol.% graphite. As a result, no information is available on the advantage of these materials.

At the same time, it is known from literature data [5-17] that phosphorus exerts a favorable influence on the physicomechanical and friction properties of metals and iron-base alloys. Thus, it was established [5-7, 9] that additions of red phosphorus or ferrophosphorus powders substantially improve the mechanical properties and wear resistance of sintered iron-powder materials. Data showing the effect of phosphorus additions on the mechanical properties of sintered iron are presented in Table 1 [6]. It was also found that a steady increase of wear resistance, accompanied by a decrease of ductility, is exhibited by sintered iron-base specimens when their phosphorus content is increased up to 3%.

The effect of phosphorus content on the wear resistance of porous iron-graphite materials was studied in [15]. Experimental results showed a large increase in the wear resistance of these materials when the phosphorus content of the charge was raised to 0.7%. Naeser and Burmeister [8] consider that sintered iron with

Table 1

Phosphorus content, wt.%	Yield strength kg/mm^2	Elongation %	Brinell hardness	Impact strength, kg-m/cm^2
0	13.14	6.7	30	0.28
0.4	33.3	14.7	76	0.5
0.8	56.9	18	146	0.1

improved properties may be produced by adding to it phosphorus in amount of up to 0.7-0.8%. Phosphorus additions greatly improve the sinterability of iron powder, which is due to the formation, on the surface of particles, of a eutectic, which melts at 1050°C.

Results of studies of the effect of phosphorus on the properties of friction cast irons are given in [10, 11, 13, 14]. Ol'shevskii and Larin established that, as the phosphorus content of friction cast irons is raised in the range between 0.118 and 2.01 wt.%, the coefficient of friction of these materials on tire steel increases, the stability of this coefficient on the whole section of the braking path is improved, and so is the wear resistance of the cast irons. The optimum phosphorus content of friction cast iron is 1.5-2% according to Ol'shevskii and 1.4% according to Larin.

The Institute of Materials Science has carried out an investigation of the effect of phosphorus on the friction properties of a sintered material based on alloyed iron. Phosphorus was added to the material in the form of a chemical compound in amounts of 0.7, 0.8, 0.9, and 1%. The resultant phosphorus-containing alloys were designated ZhF-7, ZhF-8, ZhF-9, and ZhF-10.

Test specimens were compacted at a pressure of 35-40 kN/cm^2 and sintered in a protective atmosphere at 1060 ± 10°C; they were held at this temperature for 1 hr, pressure during sintering being 150-200 N/cm^2. The friction characteristics of these alloys were determined in an IM-58 apparatus under conditions of dry friction on cast iron S421-40 with a hardness of 1.2-1.5 kN/mm^2, and were compared with those of the sintered material FMK-11 and the asbestos-resin composition FK-16-L, which were subjected to similar tests. The test conditions were as follows: 1) initial sliding velocity of friction surfaces during braking 15 m/sec; 2) change of velocity during a braking cycle from 15 to 0 m/sec; 3) braking time 5 sec; 4) specific braking power 400 $N\text{-}m/cm^2\text{-}sec$; 5) specific pressure was varied between 50 and 110 N/cm^2 in order to secure a constant specific power for all specimens tested; 6) a coefficient of mutual overlapping of friction surfaces of 0.7 was obtained by appropriate preparation of specimens for the tests. The results of these tests are given in Table 2.

Of the phosphorus-containing alloys, the best friction properties were shown by alloy ZhF-8, which had the highest coefficient of friction, combined with adequate wear resistance. The

Table 2

Material	Phosphorus content wt.%	f_{mean}	f_{max}	α	I, μ
ZhF-7	0.7	0.22	0.25	0.88	90
ZhF-8	0.8	0.27	0.325	0.83	105
ZhF-9	0.9	0.25	0.3	0.83	136
ZhF-10	1	0.25	0.32	0.75	166
FMK-11	—	0.29	0.34	0.85	373
FK-16-L	—	0.63	0.74	0.85	2040

Notation: $f_{mean,}$ mean coefficient of friction; $f_{max,}$ maximum coefficient of friction; α, stability coefficient $= f_{mean}/f_{max}$; I, wear of specimens in 100 cycles.

bending and shear strengths of these materials lie within permissible limits for friction materials, and are equal to 450 and 70 N/mm^2, respectively. The hardness of the sintered alloys is 1.5-2 kN/mm^2. The structure of these alloys is characterized by the presence of a uniformly distributed network of a phosphide eutectic consisting of ferrite and the chemical compound Fe_3P, with a microhardness of 5.5-6 kN/mm^2, surrounding the grains of alloyed iron with a microhardness of 4-4.4 kN/mm^2.

An analysis of the test results and specimen microstructures (Figs. 1 and 2) confirm the hypothesis put forward in [12, 14] for the mechanism of the beneficial influence of phosphorus on fric-

Fig. 1. Microstructure of alloy ZhF-8 (600 ×).

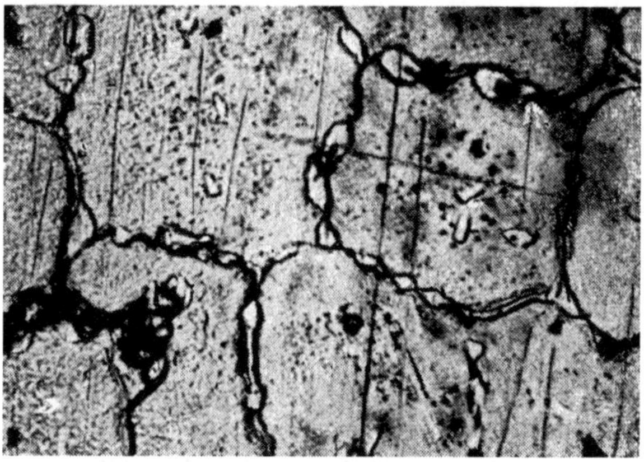

Fig. 2. Microstructure of alloy ZhF-8 (1350 ×).

tion properties. According to Kragel'skii, phosphorus promotes the formation of a cellular, heterogeneous structure of increased resistance to plastic deformation.

Studies of various phosphorus cast irons confirmed the superiority of the grade with a cellular distribution of the phosphide eutectic above those grades in which the phosphide eutectic is in the form of separate inclusions. The enrichment in phosphorus of the surface film forming during friction imparts to the film a high resistance against seizure, which secures normal operation of friction pairs.

Thus, the results of this investigation show that, by adding phosphorus to a sintered iron-base alloy, it is possible to obtain a friction material whose wear resistance is 3.5 times higher than that of FMK-11 and 20 times higher than that of FK-16-L, and which has a satisfactory value of the coefficient of friction.

Part II

In Part I it was shown that the introduction of phosphorus into sintered iron-powder materials permits us to develop friction material with a high performance and excellent physicomechanical properties. An important factor is the investigation of phenomena affecting the development of the properties of materials and the method of introducing the phosphorus into the material. The literature contains several references to methods of introducing phos-

Table 3. Results of Tests on Specimens of ZhF-8 and MK-2 on the IM-58 Friction Apparatus

Material	f_{mean}	μ	α
ZhF-8	0.27	105	0.83
MK-2	0.27	105	0.8

Test cycle: 1) Initial rubbing speed of the friction surfaces during braking, 15 m/sec; 2) change in speed during braking, from 15 m/sec to 0; 3) braking time, 5 sec; 4) specific braking power, 40 kgm/cm-sec; 5) specific pressure, 100-130 N/cm^2.

phorus to sintered materials and also their physicomechanical properties. Pugina and Shamrai [17] state that the addition of ferrophosphorus affects the physicomechanical properties of the iron more than the introduction of pure phosphorus. Reports on the effect of the method of adding phosphorus on the formation of the structure and the operating properties of sintered materials in general, and friction properties in particular, are not to be found in the literature.

We shall now deal with the results of an investigation into the distribution of phosphorus into the structure of sintered materials based on alloyed iron, and the effect on the operating properties of the introduction of Fe_3P and phosphorus salts. We investigated two materials, ZhF-8 and MK-2, which under the same laboratory test conditions have the same friction properties (Table 3). Both materials have the same degree of alloying in the iron base with other elements and differ only in the contents of phosphorus.

In the ZhF material the phosphorus was added in amounts of 0.7-1%, normally in the form of ferrophosphorus containing 15.6% P, which corresponds to the stoichiometric composition of Fe_3P. In the MK material, containing from 1 to 3.5% phosphorus, the element is added to the batch, as a rule, in the form of the salt $(NH_4)_2HPO_4$. The total average phosphorus content of the material was controlled by chemical analysis. Specimens of the ZhF and MK materials were sintered in an argon atmosphere at 1060° ± 10°C for an hour under a pressure of 150-200 N/mm^2. Both materials showed better friction properties than all the materials marked ZhF and MK under test. Alloy ZhF-8 contained 0.8% P and alloy MK-2 – 2.5%.

Then detailed metallographic analyses of the alloys ZhF-8 and MK-2 were carried out. The microstructure of alloys ZhF-8

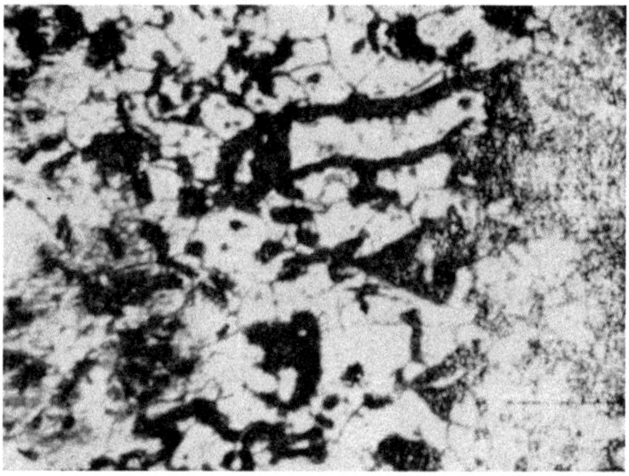

Fig. 3. Microstructure of alloy MK-2 (600 ×).

(Fig. 1) is characterized by a clear network over the grain boundary of the alloyed ferrite. The network consists of two regions varying in color, one light gray and one dark gray. Microscopic investigation of the microstructure of alloy ZhF-8 shows that this network consists mainly of the light gray region. The microhardness of the alloyed ferrite in the alloy ZhF-8 was 4000-4400 N/mm^2.

The microstructure of alloy MK-2 (Fig. 3) consists of grains of alloyed ferrite with a microhardness of 5000-5500 N/mm^2. The network over the grain boundaries is absent in this case.

To explain the chemical composition of the structural constituents noted we carried out an x-ray analysis on the electron beam of the "Sameca" type. The design of the beam equipment and its technical characteristics are given in papers [18-20]. The dimensions of the beam are 1-2μ, the accuracy of the tests is ±5%. To exclude the influence of the absorption of characteristic rays in the specimen and secondary fluorescence excitation when determining phosphorus in alloys, we used a standard homogenized alloy based on iron containing 13.8% P.

The investigation of the results of the x-ray analysis in terms of the relative P content in the structural constituents noted in the microstructure of both alloys (Table 4) suggest the following conclusions.

1. Upon addition of phosphorus to sintered friction alloys of the type ZhF in amounts of 0.8% one obtains a structure character-

Table 4. Results of X-ray Analysis of the Phosphorus Content of Structural Constituents of the Alloys

Material	Structural consituents	Phosphorus content in structural constituents, wt.%	Suggested phase composition
ZhF-8	a. Alloyed ferrite	0.8	Solid solution of P in Fe
	b. Light-gray region of network	10.8	Eutectic α = Fe + Fe_3P
	c. Dark-gray region	8.35	Pre-eutectic alloy α = Fe + Fe_3P
MK-2	a. Alloyed ferrite	1.61	Solid solution of P in Fe

ized by the formation on the grain boundaries of an alloyed ferrite of eutectic composition (Fig. 1). In this case the phosphorus added to the alloy takes part in the formation of the network, partially alloys the ferrite grains, and contributes to the creation of optimum working properties.

2. The introduction into the alloy of the more active and fusible pure phosphorus formed during the decomposition of the salt $(NH_4)_2HPO_4$ leads, as a rule, to the alloying of the metallic base of the alloy without the formation of a network over the grain boundaries of the ferrite (Fig. 3).

3. The presence in the structure of a network of eutectic composition contributes to the formation of a heterogeneous structure in the friction alloy ZhF-8, ensuring the production of optimum friction properties with a lower phosphorus content. To obtain the same friction properties with alloy MK-2 requires double the amount of phosphorus. In this case the optimum friction properties in the alloy are obtained on account of the significant strengthening in the ferrite grains.

4. For the manufacture of sintered friction alloys based on an iron–phosphorus base it is desirable to introduce the phosphorus not in the form of the salt $(NH_4)_2HPO_4$, but in the form of ferrophosphorus. This method is simpler, more economical, and more suitable and safer for the workers' health.

References

1. Strobel, Technik., Vol. 12, p. 819 (1957).
2. US Patent No. 2,849,957.
3. French Patent No. 1,145,018 (1957).
4 Japanese Patent No. 8781/58.
5. F. Eisenkolb, Planseeberichte für Pulvermetallurgie, Vol. 2, No. 1, p. 2 (1954).
6. F. Eisenkolb, F.-Wiss. Z. T. H. Dresden, Vol. 2, No. 2, p. 251 (1952-1953).
7. F. Eisenkolb. Arch. Eisenhüttenwesen, Vol. 24, p. 257 (1953).
8. G Naeser and H. Burmeister, Arch. Eisenhüttenwesen, Vol. 24, p. 251 (1953).
9. F. Eisenkolb. Powder Metallurgy [Russian translation], Moscow, Metallurgizdat (1959).
10. A. A. Ol'shevskii, Liteinoe Proizv., Vol. 7, p. 14 (1957).
11. A. A. Ol'shevskii, Investigation of Braking Railroad Cars with the Aim of Improving Their Performance [in Russian], Dnepropetrovsk (1959).
12. I. V, Kragel'skii, Proceedings of 3rd All-Union Conference on Friction and Wear [in Russian], (1960), Vol. 2, p. 262.
13. V M Gudchenko and A. M. Petrunin, "Raising the effectiveness of braking devices. Properties of friction materials," Sb. IMASh Akad. Nauk SSSR, p. 110 (1959)
14. G. V. Larin, V. P. Devyatkin, and A. Ya. Tarasenko, Proceedings of 3rd All-Union Conference on Friction and Wear [in Russian] (1960), Vol. 2, p. 278.
15. I. S. Moshkov, Vestn. Mashinostr., Vol. 10, p. 72 (1960).
16. V. Ya. Bülanov and A. D. Moshkov, TsITÉIN, No M-60-295/11, Moscow (1960).
17. L. I. Pugina and F. I. Shamrai, in: Powder Metallurgy in Machine and Instrument Construction [in Russian], Kiev, Mashgiz (1961).
18. R. Casting, Application des sondes électroniques à une méthode d'analyse ponctuelle chimique et cristallographie. Thésis Univ. Paris, ONERA, Publ. No. 55 (1951).
19. S. B Maslenkov and T. V. Egorina, Zav. Lab., No. 12 (1962).
20. K. Swobold, R. Blöck, and Plöcinger, Berg- und Hüttenmännische Monatshefte, Vol. 11, pp. 391-401 (1963).

ANTIFRICTION MATERIALS

Chapter 14

The Bearing Performance of Sintered Metal Bearings

G. C. Pratt

Glacier Metal Co.
England

Introduction

Powder metallurgical techniques are used in the manufacture of a wide range of plain bearing types, which are used in a wide variety of applications. The vast majority of these bearings are based on copper alloys.

It is convenient to deal with the types of bearings which are based on or incorporate powder metals under five main headings.

1. Steel-backed materials with a compacted nonporous sintered lining for engine bearings and wrapped bushes.
2. Steel-backed materials with a porous sintered bronze lining impregnated with plastic to form a material suitable for dry operation or operation with only a limited lubricant supply.
3. Unbacked porous sintered metal parts impregnated with oil to form self-lubricating bushes or washers.
4. Unbacked nonporous sintered metal parts incorporating graphite for operation under poor lubrication conditions.
5. Sintered polytetrafluoroethylene (PTFE) parts incorporating powder metals for nonlubricated applications.

Of these five types of bearings (1) and (3) have been established for many years and sell in very large volumes, together in the region of 10^7-10^8 parts per day. The other three types sell in much smaller quantities – together approximately 10^5 parts per day – in the specialized applications for which they were designed.

The Evaluation of Bearing Performance

The evaluation of bearing performance may be carried out at several levels:

1. On simple rigs such as the pin-and-ring machine. Usually the specimen under test is flat to avoid the necessity of forming into a bearing, and test conditions of sufficient severity are chosen to enable an evaluation to be made after a fairly short test duration of a few hours.
2. On more sophisticated rigs which simulate reasonably closely a range of practical bearing conditions. The speciman is an actual bearing and extended test durations are usually used.
3. On the actual engine or equipment in which the bearing is expected to operate, but with the equipment operating under controlled conditions.
4. On the actual engine or equipment operated under normal service conditions.

These approaches are in increasing order of certainty that the results obtained from the investigation will apply in practice, but this increased certainty carries with it the penalty of increased time before the results are available. If the field tests of category (4) are used it may be several years before meaningful measurements can be made. At the other extreme, the doubt attached to the simple accelerated tests of category (1) is such that a choice of bearing material based on such a test alone could be a wrong one. Prior correlation of the results from the simple test with a more sophisticated test or with actual practice has to be obtained, and obtained moreover for each class of material investigated.

In the author's opinion the most useful type of rig is that of category (2), and it is on this type of rig that many of the results described in this paper have been obtained. Rigs of this sort can be made sufficiently flexible for a wide range of practical applications to be closely simulated, and with a careful experimental technique results of good repeatability can be obtained. Although the bearing conditions may be slightly accelerated to enable the results to be obtained in a reasonable time, they should not be accelerated to the extent that the conditions bear no relation to practical conditions, for there is then the danger that the results will

Fig. 1. The bearing fatigue test rig. The rig simulates the automotive engine bearing condition.

bear no relation to practical results. This is particulary true when the surface bearing properties of a material are being assessed, less true when it is the fatigue strength.

For the development and evaluation of engine bearing materials a dynamic load must be applied to the bearing, through an oil film, and one of the rigs which does this is shown in Fig. 1.

The rig, which is described in detail by Duckworth and Walter [1], has eccentric weights on the test shafts which during rotation of the shaft apply a dynamic load alternately to the bearings in two pairs of test stations. The rig has two test shafts, so that a total of sixteen half bearings is evaluated in one experiment. Bearing-to-bearing variability can thus be assessed. Speed of shaft rotation is 4000 rpm, and loads of up to 6000 psi, measured by strain gauges on the connecting rods, can be applied to the test bearings.

Shaft diameter is 2.074 in. and the bearing length 1.160 in. Oil is supplied to the bearings via the connecting rods.

The loads producing fatique on the test bearings of this rig are rather lower than fatigue loads quoted elsewhere. This is because the rig is specifically designed to reproduce the edge loading normally obtained on an automotive engine bearing, arising from flexure of the crankshaft.

A very different rig is required to evaluate the performance of bearing materials under conditions of marginal or boundary lubrication or nonlubricated conditions. Here, it is not the resistance of the bearing material to dynamic loads applied through an oil film which is important, but the compatibility of two surfaces in rubbing contact, the wear they produce on one another, and the friction between them.

A rig used for this type of investigation is shown in Fig. 2. The test bearing is a bush of diameter in the range $\frac{3}{8}$-1 in., and this is placed on a test shaft capable of rotating at a range of speeds from 250 to 6000 rpm. The load is applied hydraulically upward as in Fig. 2., or, in another rig, downward by dead weights. In the former case, loads up to 1000 lb can be applied, in the latter loads up to 400 lb.

For marginally lubricated tests two techniques have been developed. In the first the load is built up in stages from 400 to 2000 psi, the maximum PV being 500,000 psi-ft/min. The bearing is drip lubricated, and the test is run on a stop-start cycle. During the 9-min stop period the oil is squeezed out from between the shaft and the bush, and at the beginning of the start period a small amount of wear takes place before a hydrodynamic film is built up. If the bearing survives to the maximum load, the test is continued for 336 hours, when the amount of wear suffered is measured.

The alternative technique is simply to initially grease a bush and run it under a moderate PV of 20,000 (150 psi) until seizure takes place. Provided the grease is carefully metered, reasonably reproducible results can be obtained.

Very good reproducibility can be obtained during the unlubricated bush testing of dry bearing materials, since one source of variability, the lubricant, has been removed. Tests are carried out over a wide range of speeds, loads, temperatures, shaft mater-

Fig. 2. The bush test rig, for studying the performance of bearing materials under boundary lubrication conditions.

ials, and shaft finishes, and the results can be used to predict performance in applications where these factors are approximately known.

Other rigs are available for the investigation of oscillating conditions, the behavior of thrust washers, and for studying the performance of sliding bearings under high loads.

Steel-Backed Materials with a Compacted Lining

Manufacture

Engine bearings are lined with three main classes of materials: lead- or tin-based white metal (Babbit), copper-based alloys, and aluminum-based alloys.

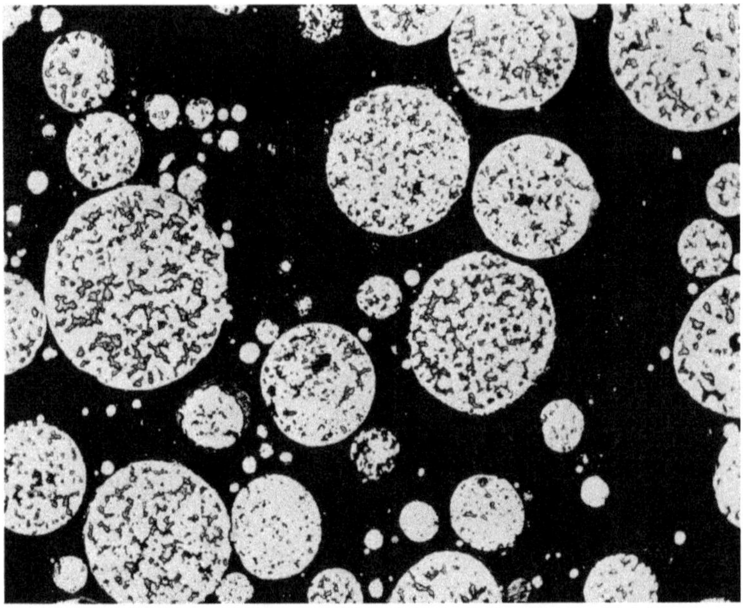

Fig. 3. A microsection of gas-atomized copper-lead particles, showing lead finely distributed within the copper. 200 ×.

For the majority of modern engines, white metal, even when the lining is thin, has insufficient strength to carry the load, and the copper-based materials such as copper-lead or lead bronze, or the aluminum alloys such as aluminum-tin are preferred. The copper-based alloys can be applied to the steel strip from which the bearings are subsequently formed in two ways: they are either cast on or applied by a powder metallurgy technique. The powder metallurgy technique competes successfully with the casting process economically and offers some technical advantages. It is used by a major plain bearing company in the United Kingdom, and two of the major companies in the United States.

The first stage in the manufacture of the bearings is to blow the copper or lead bronze powder, an example of gas-atomized copper-lead powder being shown in Fig. 3. The alloy composition is established in the melt and the liquid steam from the melt is atomized by a blast of nitrogen with a small hydrogen content. During the rapid cooling of each individual particle a fine distribution of lead throughout the particle is obtained.

In the next stage the powder is spread on to copper-plated steel strip, for typical lining thickness in the range 0.015 in.-0.060 in., the -36 B.S. powder fraction being used. The powdered strip then passes in a reducing nitrogen/hydrogen atmosphere through sintering furnaces at a temperature 800-850°C and the powder particles sinter to one another and to the copper-plated steel (Fig. 4). The "first sinter" strip then passes through a rolling mill and the lining is given a calculated reduction to eliminate porosity. Re-sintering and possibly a final size roll completes the process.

A typical copper-lead lining obtained in this way is shown in Fig. 5. This may be contrasted with gravity and centrifugally cast copper-lead structures shown in Figs. 6 and 7. It will be seen that one advantage of the sinter process is the finer structure and lead distribution that is obtained.

In the United States a popular material for less heavily loaded engines combines casting and powder metallurgy techniques. A sintered layer of 40% Ni, 60% Cu is first applied to the steel and this layer is then infiltrated with white metal to leave a 0.003 in. thick layer of soft bearing material above the hard cupronickel.

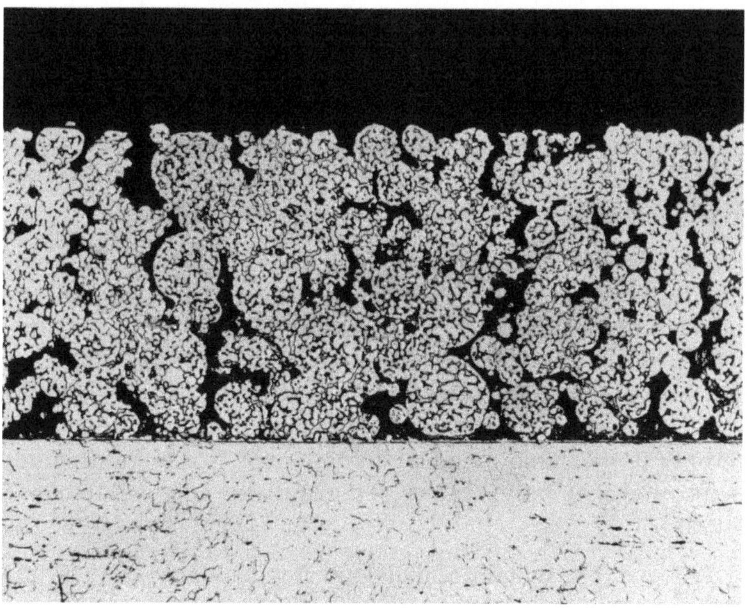

Fig. 4. Copper-lead powder on steel after the first sintering operation. 100 ×.

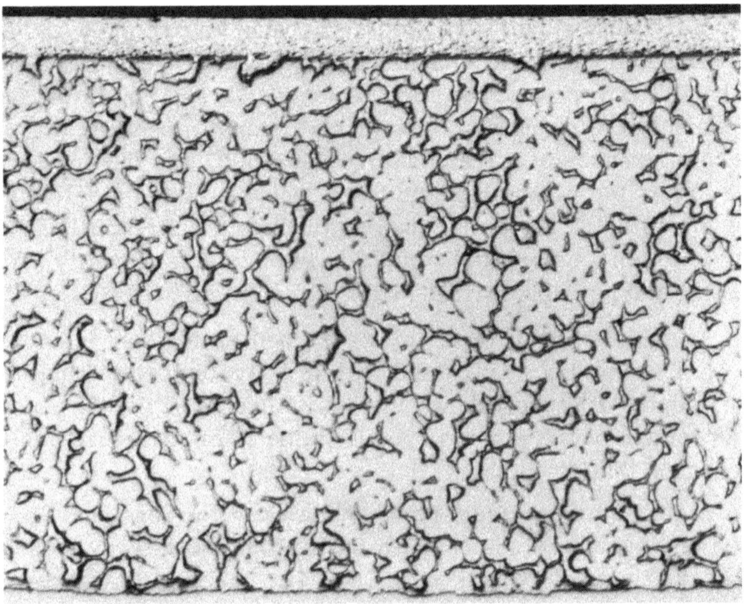

Fig. 5. A sintered copper-lead bearing lining, overlay plated. 300 ×.

Composition Properties, and Performance

As in all metal-based bearing materials, the requirements of high fatigue strength and good surface properties cannot be combined in one material. A composition must be chosen which in the light of the operating conditions of the bearing offers the best compromise between these opposing requirements.

The material referred to above, which consists of a lead-base white metal cast into a sintered cupronickel lining, has excellent surface properties, since approximately 0.003 in. of white metal lies above the cupronickel in the final bearing. White metal (Babbit) has good conformability to slight deviations in shaft contour or aligment; it embeds dirt well, thus producing virtually no shaft wear; it shows no tendency to "pick up" (local welding onto the shaft); and it is not prone to seizure under adverse lubrication conditions.

However, the dynamic load-carrying capacity of white metal is not high, and this is reflected in the fatigue rating of the impregnated cupronickel material. Table 1 shows that it is similar to that of a 0.003 in.-thick white metal on steel.

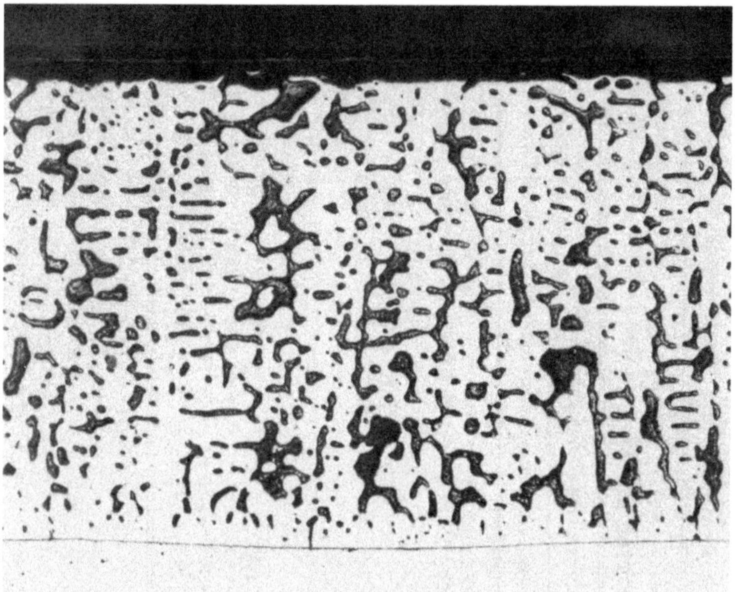

Fig. 6. The structure of a gravity-cast copper-lead lining. 200 x.

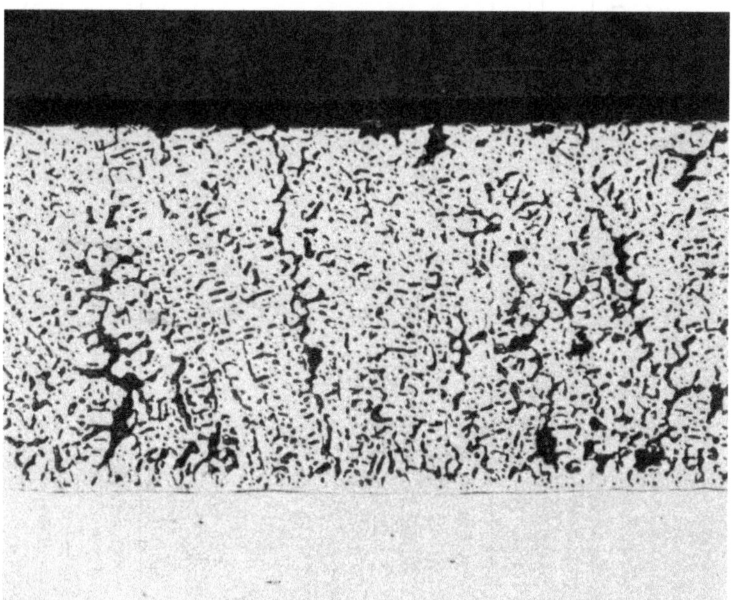

Fig. 7. The structure of a centrifugally cast copper-lead lining. 200 x.

Table 1. Fatigue Strength of Bearing Materials

Material	Lining thickness, in.	Composition					Fatigue strength of lining, psi*
		copper	lead	tin	antimony	aluminum	
White metal	0.015	4		87	9		1800
White metal	0.003	4		87	9		2200
0.003 in. white metal over sintered cupronickel							2400
Copper-lead (sinter)	0.015	60	40				3000
Overlay-plated copper-lead (sinter)	0.015	70	30				3400
Overlay-plated lead bronze (cast)	0.015	75	23.5	1.5			3700
Overlay-plated lead bronze (sinter)	0.015	74	22	4			5000
Aluminum-tin	0.015	1		20		79	4300

*Overlay fatigue may take place at lower load.

Table 2. The Tensile Strength and Hardness of Sintered Copper-Based Bearing Linings

Composition, wt. %			Average U.T.S., psi	Average hardness, HV
copper	lead	tin		
60	40		11,000	32
70	30		12,500	37
74	22	4	17,000	50
80	10	10	26,000	70

The Copper-Leads

A material which achieves a useful improvement in strength without sacrificing to a great extent its surface properties is a high-lead copper-lead, with a lead content in the region of 40–45% by weight. Fatigue strength is about 50% greater than that of a similar thickness of white metal lining, yet the surface properties are sufficiently good for the material to be used in automotive engines without an overlay. In the United States large quantities of this type of bearing known as "Intermediate" are used in passenger car engines.

A rather stronger material contains approximately 30% lead. The surface properties of this material are less good and as an engine bearing this material is usually operated with a thin lead-tin or similar overlay. The surface properties or antiseizure properties of an overlay of this sort are very good. Since it is thin, of the order of 0.001 in. thick, fatigue resistance is almost as high as that of the underlying copper-lead, and the overlay because of its high tin content resists corrosion and prevents attack of the lead in the lining by the engine oil.

The mechanical properties of the 70/30 composition are included in Table 2, and its fatigue rating in Table 1.

Fatigue of copper-lead can occur under severe loading conditions, both in sintered and cast material. The pattern of fatigue cracking is similar to that in white metal: vertical cracks from the surface, turning and running parallel to the steel backing a little way above it. Figure 8 shows a typical fatigue crack in sintered copper-lead. The bearing which suffered this fatigue had

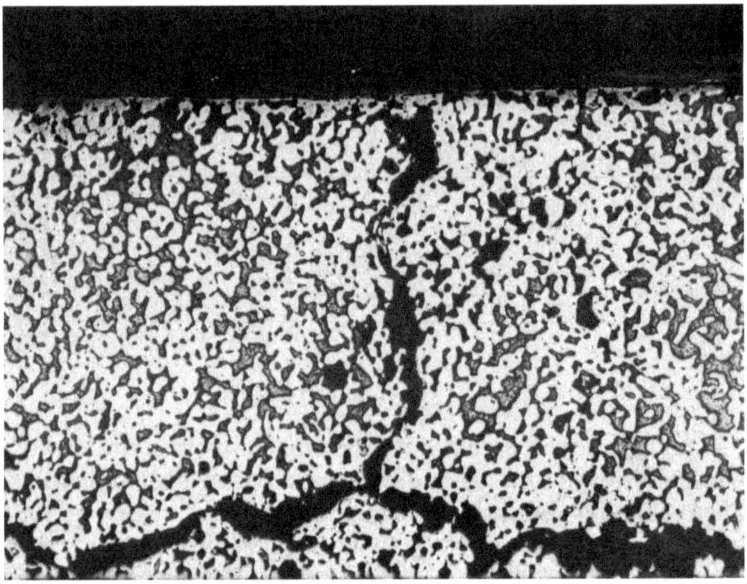

Fig. 8. A fatigue crack in a sintered copper-lead bearing lining. 160 ×.

Fig. 9. Overlay-plated sintered copper-lead bearings after 226,000 miles in a diesel engine.

completed in fact only 70,000 miles in a locomotive diesel engine; fatigue occurred because of heavy edge loading arising from shaft misalignment. More typical of the performance of sintered copper-lead in a diesel engine is Fig. 9, which shows a pair of bearings in near perfect condition after 226,000 road miles plus 3600 pumping hours in a tanker vehicle. The only damage to the bearings in this time was a light scoring of the overlay by dirt.

The Lead-Bronzes

If a dynamic load-carrying capacity beyond that of a straight copper-lead is required, lead content can be reduced still further to around 20%, and a small tin addition to strengthen the copper matrix made. The 74% copper, 22% lead, 4% tin composition shown in Tables 1 and 2 is typical of this class of lead bronze. Hardness of the lining is in the range 40-60 HV and fatigue strength is increased to 2.5-3 times that of white metal. Again an overlay is required before composition of this sort can safely be operated in an engine.

This type of tin bronze is also used extensively as a bush material, the steel-backed sinter being wrapped to form a wide range of bush sizes from 3/8 in. to 8 in. bore. It is common prac-

Table 3. Bush Performance under Boundary Lubrication Conditions (Specimen size 5/8 in. bore × 3/4 in.; load increased from 400 to 2000 psi or until seizure obtained; shaft speed 1500 rpm, stop-start cycle, oil lubricated)

Material	Maximum PV attained, psi - ft/min*	Hours run at maximum PV before seizure
80/10/10 lead bronze	200,000	7
74/22/4 lead bronze	300,000	0.5
80/10/10 lead bronze, indented and graphited	400,000	1
8% graphite in tin bronze	300,000	3
Oil-impregnated porous bronze	300,000	1
Acetal copolymer sinter-bonded to steel	500,000	336†

*Bushes run 24 hours at 100,000, 24 hours at 200,000 and so on until seizure obtained.

†Final wear less than 0.001 in.

tice to groove the bush bore, to assist the supply of lubricant from the oil hole to the bearing surface, the groove pattern depending on the nature of the lubricant – grease or oil – and whether a pressure or gravity feed is used. Applications cover a very wide field, but the following are typical: main and small end bushes in low-duty compressors, king pin bushes, rocker and cam follower bushes, gear box bushes, and machinery linkages.

Certain applications call for a still higher degree of hardness and strength than can be obtained with the type of lead bronze discussed above. For these applications the lead content must be again reduced to the order of 10%, and tin content increased to the same level. Normal hardness of the 80/10/10 composition is in the range 60-80 HV, but it may be increased by cold working to as much as 130 HV.

Because of its tendency to seizure under poor lubrication conditions the material is normally used with grease lubrication rather than oil and run against a hardened mating surface. The material does not find acceptance as an engine bearing material. Applications are largely confined to the automotive field, small end bushes, and gear box bushes in particular.

The surface bearing properties of the 80/10/10 composition may be substantially improved by incorporating in the surface of the material a proportion of solid lubricant such as graphite. The surface of the strip is provided in a rolling mill with a pattern of indents which are subsequently filled with a graphite/nylon composition. Under conditions of sparse oil lubrication a big improvement in performance can be obtained in this way, and results illustrating this improvement are shown in Table 3.

In another material with somewhat similar bearing properties, graphite is distributed throughout a tin bronze structure by incorporating a proportion of graphite with the copper and tin powders, which in this case are post-alloyed during the sintering process. Although the resultant structure is rather weaker, because of the graphite constituent, than the 80/10/10 material, the fine even distribution of the graphite (Fig. 10) results in a high level of performance under adverse lubrication conditions, relative to more conventional materials.

A successful application has been the primary gear bushes in an automotive engine where the design permitted only restricted

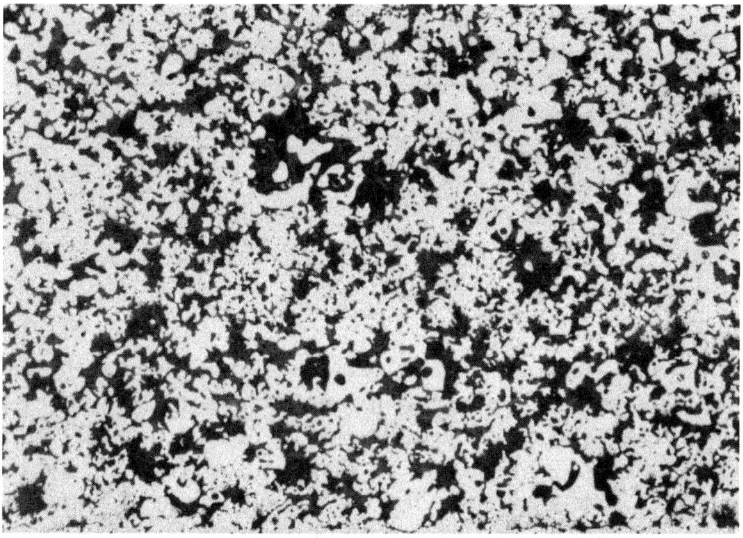

Fig. 10. Microsection of a graphited bronze bearing lining. 8% graphite in tin-bronze. 300 ×.

access of mist lubrication to the bushes. The material has also proved successful as a motorcycle big end bearing.

Steel-Backed Materials with a Porous Plastic-Impregnated Lining

Manufacture

There are two types of bearing material in this class, both originating in the United Kingdom. One, in which the impregnant is a mixture of PTFE and lead, is a dry bearing material; that is, it is primarily intended for operation in the complete absence of lubricant. The other is impregnated with the thermoplastic acetal copolymer and is intended for operation with a minimum of lubrication; for example, the bearing might be lubricated on assembly only.

Both these materials consist of copper-plated steel strip to which is sintered a layer of tin bronze approximately 0.010 in. thick. To ensure maximum porosity for impregnation while retaining a high degree of strength (Table 4) a narrow particle size fraction is used which may be -60 + 100, -85 + 120, or -120 + 200 B.S. mesh.

Table 4. The Strength of Porous Bronze Sinter (89/11 tin-bronze -60 + 100 B.S. fraction 0.040-0.095 -in.-thick layers sintered onto steel; steel ground off before test)

Porosity, %*	Average tensile strength, psi
50-53	2860
46-49	3660
43-46	4250
39-42	5170

*Includes surface porosity.

Into this layer the PTFE/lead is roll-impregnated in the form of a thick paste, the impregnation conditions being adjusted to leave a very thin layer of PTFE/lead above the bronze. The PTFE is then sintered above its transition temperature of 327°C, and the strip given a final size roll. During the running in of bearings of this sort, the thin PTFE/lead layer is removed to expose in the bearing surface a small proportion of bronze. It is the very high

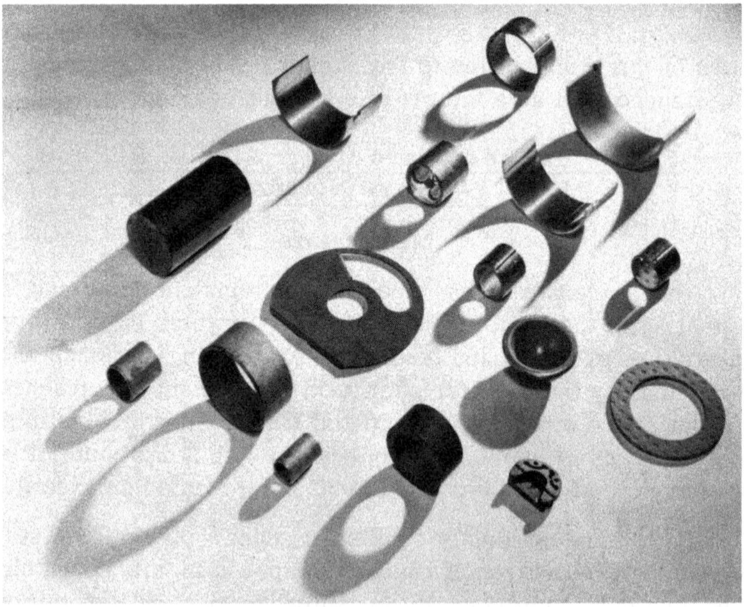

Fig. 11. A range of bearings, based on powder metals, manufactured by the Bearing Products Division of Associated Engineering Ltd., England.

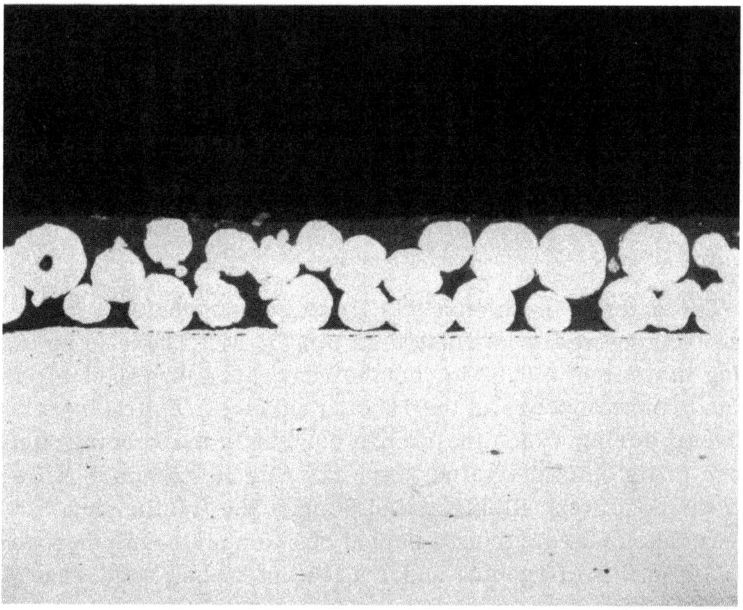

Fig. 12. The structure of PTFE/lead-impregnated porous bronze, a bearing material for operation without lubrication. 100 ×.

degree of dry wear resistance of this PTFE/lead/bronze surface which is the particular virtue of the material.

In the second material, impregnation of the acetal copolymer into the bronze is carried out in such a way to leave a thick (0.010 in.) machinable layer above the bronze, and the bronze acts therefore simply as a key via which the plastic is attached to the steel backing.

In one form of this bearing the plastic surface is provided with a pattern of indents which act as reservoirs for the small quantities of grease which this bearing requires for efficient operation. Examples of this and of the PTFE-based bearing are shown in Fig. 11, and typical microstructures are shown in Figs. 12 and 13.

PTFE/Lead-Impregnated Porous Bronze

Properties and Performance

During operation, a dry bearing – a bearing capable of operating without lubricant – wears at a finite rate which is determined most strongly by load, rubbing speed, and ambient temperature, and also by such factors as shaft material and roughness. Information on all these factors must be available before the wear rate of the bearing, in other words its life in the application, can be predicted. It must also be known whether the load is stationary or rotating relative to the bearing, whether operation is continuous or intermittent, and whether excessive dirt contamination is likely to occur.

The manner in which a PTFE/lead-impregnated porous bronze bearing wears is illustrated in Fig. 14. There is an initial bedding in wear of 0.0005 in., but the wear rate drops suddenly as soon as bronze appears in the rubbing surface. For the next 0.001 in. of wear, during which the bronze content in the rubbing surface is close to its optimum value, the wear rate is extremely slow. Under the conditions illustrated of 80 psi, 250 ft/min, this 0.001-in. wear takes place during a period of 600 hours. In other words, after 1700 miles of rubbing under a load of 80 psi, total wear was only 0.0015 in.

Fig. 13. The structure of sintered bonded acetal copolymer showing one edge of an indent. The indents provide a series of grease reservoirs in the bearing surface. 100 ×.

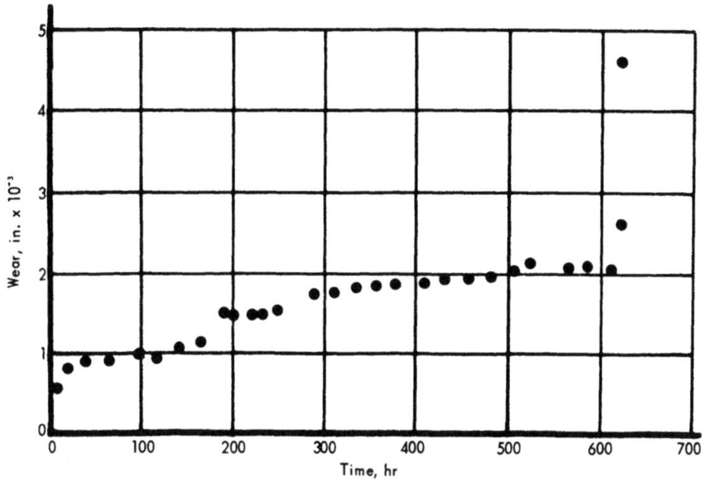

Fig. 14. The wear behavior of PTFE/lead-impregnated porous bronze. 20,000 PV; continuous rotation; load stationary relative to the bearing.

Toward the end of the life of the bearing the bronze content in the rubbing surface begins to rise above its optimum value. As more and more bronze is exposed in the rubbing surface, wear increases at an increasing rate, and the bearing has reached the end of its useful life.

The most useful parameter for determining the wear behavior of a dry bearing is PV, the product of the bearing load in psi and the rubbing speed in ft/min. The life of a PTFE/lead-impregnated porous bronze bearing is determined to a close approximation by the product PV and only minor differences are obtained between life at high load and low speed and high speed and low load, provided the product PV is constant.

Figure 15 shows the relationship between PV and the life of PTFE/lead-impregnated porous bronze bushes, the bushes running continuously against mild steel shafts, and the load being stationary relative to the bush. Life ranges from 4000 hours at 5000 PV down to 200 hours at 50,000 PV.

Under more favorable conditions, that is, intermittent operation such that the temperature build-up usually associated with high-PV operation does not occur, life at 50,000 PV can be extended to 500 hours. Life is also extended by a factor of two or more if the load rotates relative to the bush so that wear takes place over the whole of the bearing surface area and not just over one half of it.

Under less favorable conditions such as a dirty or high-temperature environment lower lives will be obtained. For instance, life is reduced by a factor of two if the bush is operating in an ambient temperature of 120°C, by a factor of four if the temperature is 200°C. Maximum operating temperature is 250-300°C. A wide range of shaft materials are compatible with this type of bearing material: all steels including stainless, chromium plate, anodized aluminum, etc. It is essential, however, that the mating surface be provided with a good finish of less than 16 microinches if optimum performance is to be obtained.

The friction coefficient of PTFE/lead-impregnated porous bronze, is, in the majority of applications, in the range 0.10-0.20. At high loads, however, the friction may fall below 0.10, and when the load is in excess of 1000 psi, values of 0.05 or less can be obtained. This feature of the material makes it particularly suitable as a sliding bearing material in civil engineering applications. It is being used, for instance, to an increasing extent in bridge bearings. Loads up to 4000 psi can be readily carried, and a friction coefficient of 0.02-0.03 obtained.

More conventional applications of the material are in the automotive industry (suspension and pedal bushes, for instance), in the aircraft industry (swirl vane bushes in jet engines), in textile and domestic machinery, and in mechanical handling, agricultural, electrical, and pumping equipment. In the latter applications the material operates quite satisfactorily immersed in water and other non-lubricating fluids.

Polyacetal-Impregnated Porous Bronze

Properties and Performance

Certain thermoplastics have surface properties which enable them to perform extremely well compared with the more conventional metal-based materials under conditions of marginal lubrication, i.e., when the lubricant supply and bearing conditions are not such that full hydrodynamic operation is established. Unfortunately, thermoplastics possess poor mechanical properties relative to

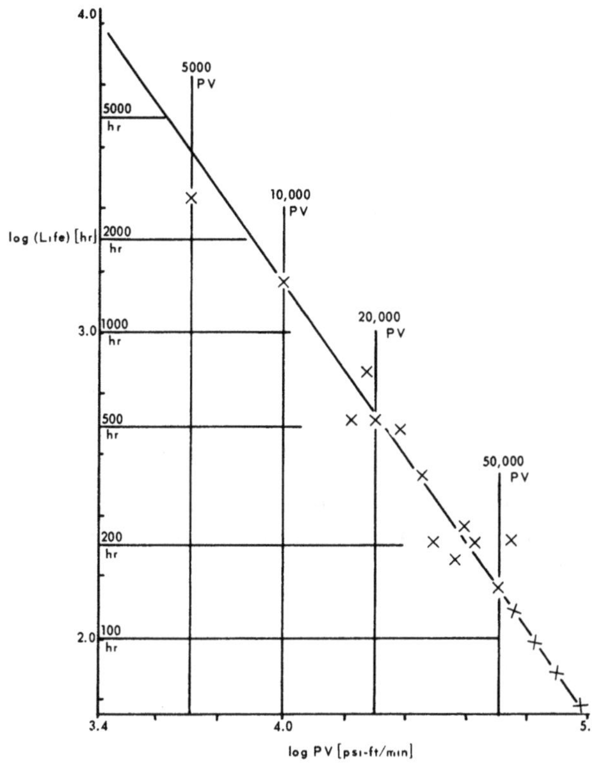

Fig. 15. The effect of PV on the life of PTFE/lead-impregnated porous bronze. $^5/_8$ in.-bore bearings; continuous rotation; load stationary relative to the bearing. Mild steel shafts ground and polished.

the bearing metals, particularly at elevated temperatures, and they also suffer from high thermal expansion and low thermal conductivity.

A way of overcoming these defects, and making use of the excellent bearing properties of the thermoplastic surface, is to bond the thermoplastic to steel using an interlayer of porous bronze sinter to key the plastic to the steel backing.

In this way a material with an exceptional combination of load-carrying capacity and surface properties can be made. The deformation under a compressive load of acetal copolymer ainter-bonded to steel is shown in Fig. 16. At room temperature no significant deformation was obtained until a load of 20 tons/in.^{2} was reached.

The bush performance of this type of bearing is shown in Fig. 17. Performance is somewhat dependent on rubbing speed, but the life obtainable from a bush grt ased on assembly, but with no subsequent lubrication, is well above that which can be obtained from other materials under these conditions. This is apparent from the results in Table 5.

The friction of an initially greased bush of this sort is low – of the order of 0.01, rising to 0.1 as the grease becomes exhausted and the bush nears the end of its life. If it is possible to relubricate the bush, even only infrequently, the friction will remain low and the bush will continue to operate indefinitely without wear.

The material has only recently appeared on the market, but has already proved successful in a number of applications. As

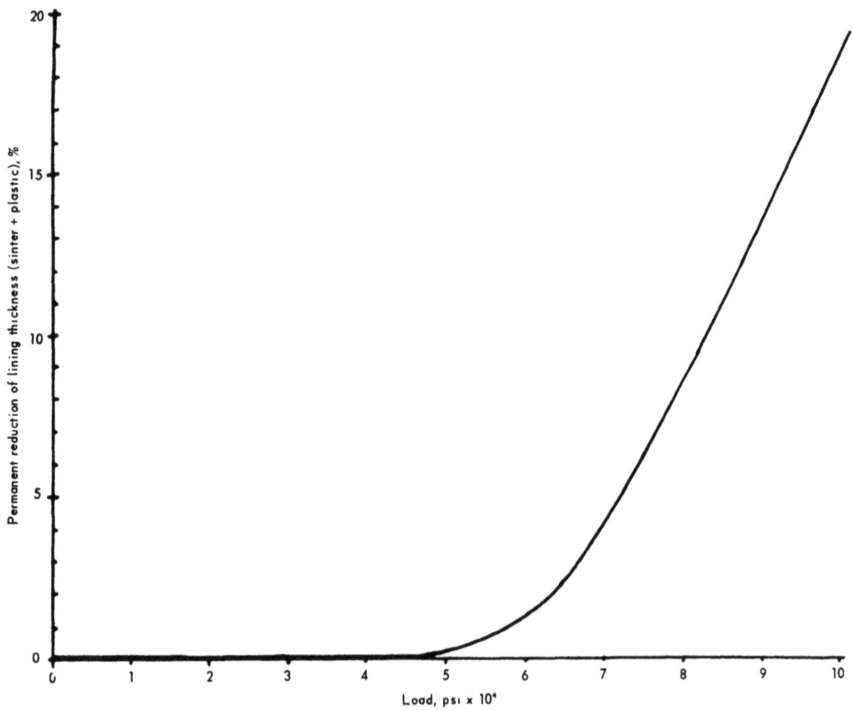

Fig. 16. The compression of acetal copolymer sinter bonded to steel. Specimens 1 in. square; lining 0.012 in.; sinter 0.010 in. plastic.

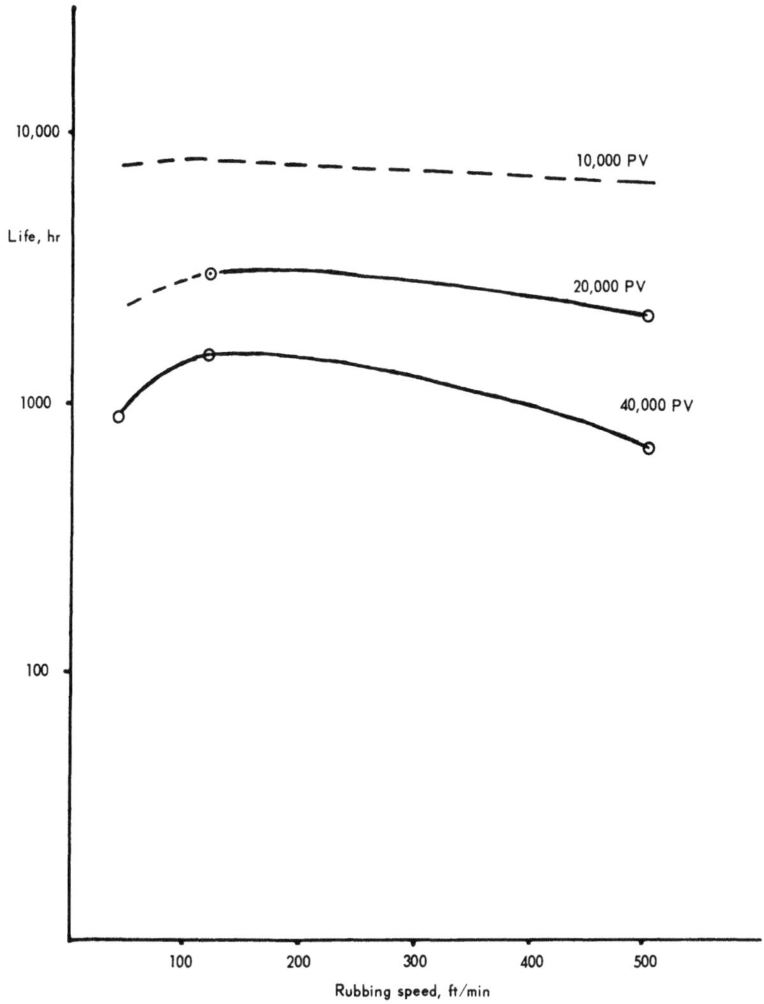

Fig. 17. The performance of acetal copolymer bonded to steel. $^5/_8$-in.-bore bearings, indented surface initially greased; no subsequent lubrication; continuous rotation; load stationary relative to the bearing. Mild steel shafts grounded and polished.

one would expect, it has shown to particular advantage in applications where the hydrodynamic oil film required by other lubricated bearing materials for successful operation is not built up, i.e., under conditions of high load or low speed, under oscillating or sliding conditions, or where repeated stopping and starting is required. In other applications the attraction has been a significant saving made possible by the elimination of a lubricant system.

Table 6. The Physical Properties of Oil-Impregnated Porous Metal Bearings

Matrix composition	Porosity, %	U.T.S., psi	Compressive strength, psi	Thermal expansion $\times 10^{-6}$, °C^{-1}
90/10 copper/tin	18-23	18,000	20,000	18
75/25 iron/copper	20-25	25,000	30,000	14
95/5 aluminum/ copper	18-20	20,000	24,000	25

bearing with an external lubricant supply. However, because of the porosity, oil can be forced from the oil film in the loaded area into the bearing wall, the oil film being replenished from the unbacked part of the bearing. There is thus a circulation of oil within the bearing itself.

The leakage of a proportion of the oil film pressure away through the pores of the metal results in a loss of load-carrying capacity of the oil film. Higher values of speed and oil viscosity are required to achieve a fluid film with porous metal bearings than with solid-wall bearings.

It is common practice to refer to the bearing capacity of oil-impregnated porous metal bearings in terms of PV, and a figure of 50,000 psi-ft/min as the limit of bearing capacity is often quoted. However, this is an unsophisticated approach, which does not take into account the regime under which the bearing is operating – whether in fact the bearing is operating under conditions of hydrodynamic or boundary lubrication. If the former is the case, a higher PV value than 50,000 may be possible; if the latter, then 50,000 PV will only be achievable if more or less frequent relubrication is acceptable. Under the test conditions of Table 5, where the shaft speed was low, a life of only 500 hcurs at 20,000 PV was obtained. Other conditions which do not favor the formation of a hydrodynamic film and under which the load-carrying capacity of oil-impregnated porous bronze is limited are oscillating conditions and frequent stopping and starting.

The criteria necessary for the establishment of a hydrodynamic oil film in porous bearings have been considered by Marshall and Morgan [3]. On the assumption that the change from boundary to hydrodynamic lubrication occurred at an eccentricity ratio of 0.8, they obtained a series of curves which enable the designer to predict whether or not a bearing will operate in the hydrodynamic condition.

One of the important parameters which determine this is the permeability of the bearing wall. Low permeability restricts the passage of oil from the oil film supporting the load into the bearing wall, and higher loads can in consequence be carried. It is advantageous therefore to make the porosity as fine as possible, while maintaining the total percentage porosity, i.e., the oil capacity of the bearing, as high as is consistent with strength and ability to maintain an interference fit with the housing.

However, if one proceeds too far in the direction of fineness of porosity there is a danger that the smearing of the bearing surface by the shaft, which will tend to occur during starting or stopping or because of accidental adverse conditions, will seal the surface porosity and cut off the supply of oil to the clearance space. This is particularly dangerous if the bearing is operating under conditions of rotating load when the whole of the surface may become sealed.

An interesting development in this respect is the advent of a porous bronze bearing with a thin film of microporous nickel on the bore. A pore size of the order of one micron gives this layer a very low permeability, and a load-carrying capacity ten times that of the conventional porous bronze bearing is claimed. Because of the resistance of the nickel to smearing the fine porosity is not sealed during operation.

A possible disadvantage of this type of bearing, however, is that the low permeability of the surface will prevent re-absorption of the oil at the bearing edges, which is part of the circulation of oil within the bearing previously mentioned, and that the oil loss from the bearing will in consequence be relatively high. Thus although the load-carrying capacity of this type of bearing will be high, this may be offset by a reduction in life.

A similar advantage of antismearing properties, making feasible reduced permeability and higher performance is claimed for porous aluminum bearings. The technical difficulties of producing this type of bearing have recently been overcome in the United States [4], and of course the use of aluminum in place of bronze offers a considerable economic attraction.

Unfortunately the high thermal expansion of aluminum tends to lead to bore closure and loss-of-interference problems with these bearings, and it may be that their use will be confined to lightly loaded applications with little heat generation within the bearings.

Unbacked Bearings Incorporating Graphite

Manufacture

This section covers unbacked bearings again made by conventional powder metallurgy techniques, but with full compaction aimed at eliminating porosity from the material. Powder metallurgy techniques are used to enable the incorporation of a proportion of a solid lubricant such as graphite.

A number of such materials are on the market. They offer very useful poperties for certain special applications, although the number sold is several orders of magnitude less than the oil-impregnated porous bearings of the previous section.

Manufacture consists of first blending the metal powder with fine natural graphite followed by compacting in hydraulic or mechanical presses. The green compacts are then sintered and finally recompacted at elevated temperature.

Composition, Properties, and Performance

The common matrix material of this type of bearing is a bronze, either a straight tin bronze or a lead bronze, but brass-, iron-, and nickel-based materials are also available. Graphite contents range from 4% to 10% by weight.

Some properties of typical compositions are given in Table 7. As one would expect, strength drops rapidly with increasing graphite content. The lower graphite grades are therefore the more suitable for severe loading conditions, the higher graphite grades being reserved for high-speed applications where the loads are not so severe. The tin bronzes are somewhat higher in strength than the lead bronzes, but the latter give a superior performance in dry and water-lubricated applications.

For very-high-temperature applications, in excess of 350°C, the iron and nickel compositions come into their own, the nickel being the more corrosion-resistant material but, of course, more expensive.

Table 7. The Physical Properties of Sintered Bearing Materials Incorporating Graphite

Matrix composition	Graphite content, wt.%	Tensile strength $\times 10^{-3}$, psi	Compressive strength $\times 10^{-3}$, psi	Radial crushing strength $\times 10^{-3}$, psi	Brinell hardness	Thermal expansion $\times 10^{6}$, °C^{-1}
Tin bronze	6	9.0	58	28	80	18
Tin bronze	8	6.7	40	22	72	17
Tin bronze	10	3.8	28	20	68	16
Lead bronze	8	4.5	28	22	62	20
Iron	6	17	71	41	65	13
Iron	8	12	39	31	57	12

Under completely dry (nonlubricated) conditions the performance of this type of material can be disappointing. At ambient temperatures up to about 250°C, the PTFE-based material described in a previous section will in general show a higher degree of resistance to wear and lower friction. However the graphited metals will operate at temperatures considerably above that which PTFE can withstand. For operation at these high temperatures the graphited metals are moreover to be preferred to the carbon matrix bearing materials, which are often in difficulty because of their low expansion coefficient, hardness, and brittleness.

With some lubrication, preferably oil lubrication, the graphited metals perform well. The result in Table 3 for a graphited tin bronze applies equally to the unbacked and the steel-backed form.

Under a high-PV stop-start oil-lubricated condition, the PV capabilities of the graphited materials were well above that of ungraphited material.

Another virtue of the graphited metals is that they seem able to survive under very adverse conditions of dirt contamination. Graphited metal bearings are operating successfully in such abrasive atmospheres as coke and coal dusts and foundry sand. The materials also seem tolerant of a fairly rough surface finish on the mating material, and the manufacturers state that a surface finish up to 60-80 microinches is acceptable.

Operation in nonlubricating fluids is often successful, and the materials are used in hot water circulating pumps and in pumps handling styrene for the chemical industry and a water-soluble oil mixture in coal cutting machinery. Other applications have been found in looms for the textile industry, motorcycle small ends, and press tool guides. There are several high-temperature applications in the steel industry and in baking ovens.

Sintered PTFE Parts Incorporating Powder Metals

Manufacture

Before concluding this review mention should be made of a type of bearing which although not strictly covered by the title nevertheless is made by powder metallurgy techniques and contains powdered material.

Polytetrafluoroethylene has been referred to in a previous section as the polymer which mixed with lead powder is impregnated into steel-backed porous bronze to make a dry bearing material – not requiring lubrication – with an exceptional degree of resistance to wear. Dry bearing materials may be made in the converse manner of putting powdered metals or other materials such as glass fiber into a PTFE matrix, and a number of such materials are on the market. These filled PTFE materials have the advantage that they can be molded into a variety of shapes and are machinable. Their load-carrying capacity, however, is not as high as that of the steel-backed material.

Although PTFE is a thermoplastic, that is, it belongs to the same family of polymers as nylon, the very high molecular weight which it is necessary for the polymer to have before it possesses a reasonable degree of mechanical strength means that even above its crystalline melting point of 327°C viscosity is very high. The polymer can in consequence not be processed by the normal thermoplastic technique of injection molding. Instead the polymer powder is compacted at high pressure and subsequently sintered, either free or under pressure, at about 360°C.

Filled PTFE compositions are made by blending the filler and the PTFE in powder form, or adding the filler to an aqueous dispersion of PTFE, coagulating the dispersion, and drying. By special techniques the filler—PTFE mixture can be made to flow sufficiently well without segregation for it to be molded on conventional powder metallurgy presses. Large numbers of parts are now made in this way at production rates approaching those of powder metal parts. Examples are shown in Fig. 11.

Composition, Properties, and Performance

Unfilled PTFE is a rather soft material with a tendency to creep under load. Although the friction of the material is exceptionally low, the wear rate of the unfilled material is very high.

Fortunately the wear rate can be reduced by the choice of a suitable filler by factors as high as 1000-10,000. The greatest degree of wear resistance is shown by the compositions containing powdered bronze as a filler. Alone, powdered bronze produces a high degree of wear resistance only at low temperature – wear rate increases as the ambient temperature is increased above room temperature or as the bearing temperature rises due to friction heat. A substantial improvement can be made by incorporating an equal proportion of graphite in the material. Thermal conductivity is thereby increased, the rise in bearing surface temperature is restricted, and a major improvement in performance is obtained.

The 20 vol.% graphite, 20 vol.% bronze composition has excellent bearing properties at low rubbing speeds (less than 100 ft/min) at PV's up to 10,000 psi-ft/min. It also performs extremely well in water-lubricated applications. At high speeds, however, its dry bearing performance falls off, and for this reason a new composition has recently been introduced which contains a small proportion of bronze combined with red lead oxide. The resistance to wear of this composition under dry conditions is very high, at all speeds. At 20,000 PV, for instance, at 250 ft/min a wear rate of only 4×10^{-6} in./hour, or 0.004 in. in 1000 hours, is obtained. However, the material is not suitable for underwater operation.

References

1. Duckworth, W. E and Walter, "Fatigue of Plain Bearings," International Conference on Fatigue of Metals, I Mech. E. (1956).
2. Morgan, V. T. and Cameron, A., "Mechanism of Lubrication in Porous Metal Bearings," Conference on Lubrication and Wear, I. Mech. E. (1957).
3. Marshal, P. R. and Morgan, V. T., "A Review of Porous Metal Bearing Development," Lubrication and Wear, Fourth Convention, I. Mech. E. (1966).
4. Storchheim, S., "Aluminium Powder Metallurgy Finally Made Commercially Practical," Progress in Powder Metallurgy, Vol. 18, (1962).

Chapter 15

Porous Metal Bearings

V. T. Morgan
Bound Brook Bearings Ltd.
England

Introduction

The first suggestion that the porous products of powder metallurgy could be used as self-lubricating bearings appears to have been made sometime in the early 1920s. The idea probably originated from attempts to overcome the heat conductivity limitation of oil-soaked wooden bearings. Fortunately, the conventional bearing bronze compositions could be produced by sintering and pressing the copper and tin powders which were then available, and called for nothing spectacular in the way of presses, sintering atmospheres, or furnace equipment. Although these early products were capable of only a fraction of the duty to which porous metal bearings are subjected today, their rapid development was possible because the techniques and principles involved in their manufacture were already available from other applications of powder metallurgy. The current (1964) world production of porous metal bearings is estimated to be about eight millions per day, and is still expanding.

In common with many other products, porous metal bearings have to compromise between a number of conflicting desirable characteristics. In order to appreciate what these are and how they arise, it is necessary to understand something of the method of manufacture and the general properties of the porous materials. As many excellent reference books on powder metallurgy are now available [1-3], only that aspect of the industry directly concerned with porous metal bearings will be described here.

Porous metal bearings are produced by powder metallurgy, which is generally employed for one or more of the following reasons:

1. The melting point of the metal or alloy is so high that fusion techniques are impracticable.
2. The desired composition cannot be achieved by casting techniques (due to segregation, for example).
3. The sintered article has special properties (such as controlled porosity).
4. Components can be fabricated more economically.

With porous metal bearings the last two (and sometimes the last three) reasons apply.

The porosity arises from the incomplete solidification of the compressed particles of metal powder, and the pores are therefore intercommunicating, as, for example, in a bed of sand.

The essential steps in the process are:

1. Metal powders are mixed to give the desired alloy composition; for example, 89% of copper powder, 10% tin powder, and 1% of powdered graphite are mixed to produce a graphited bronze.
2. The mixed powders are compacted to the desired shape in precision tools on automatic presses at pressures between 10 and 35 tons/in^2. The compacted briquettes are sufficiently strong to permit handling.
3. The "green" compacts are sintered in a reducing atmosphere. With a 90:10 bronze a temperature of about 800°C is used, and with ferrous alloys temperatures in excess of 1000°C are necessary.
4. The sintered compacts are repressed to remove distortion and to impart dimensional accuracy and a smooth surface finish.
5. The pores of the coined components are filled with lubricating oil by vacuum impregnation.

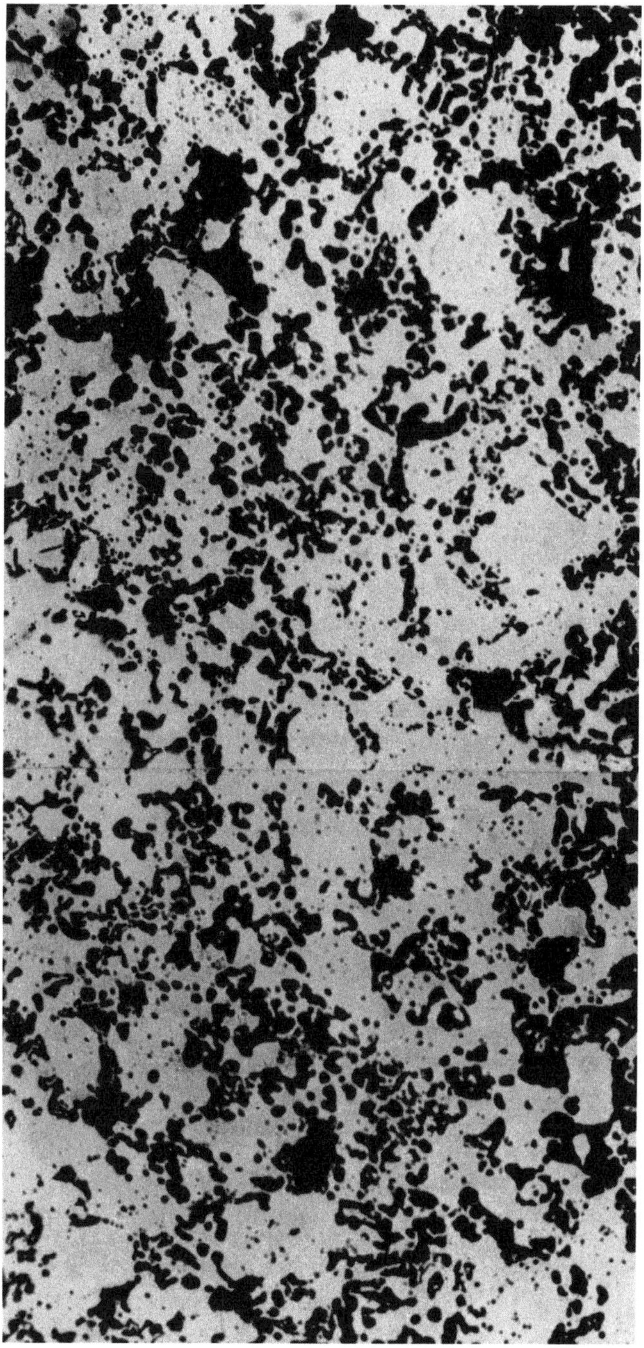

Fig. 1. Microsection of porous metal bearing. Unetched, × 100. (Dark areas are the pores.)

6. The finished bearings are packed in a nonabsorbent wrapping, to protect them from dust and avoid loss of lubricant.

Figure 1 shows a microsection of the porous structure obtained. As an alternative to oil, the pores may be impregnated with PTFE to give a dry bearing, or, where the porosity is not required and increased mechanical strength is essential, with molten metal (of lower melting point than the matrix, of course).

Physical Characteristics of Porous Metals

Porosity

The quantity of porosity is given by

$$f = 1 - \frac{\rho}{\rho_0} \tag{1}$$

where f is the fractional porosity, ρ is the apparent density, and ρ_0 is the density of the metal in the nonporous form. For a given alloy composition ρ_0 is a constant and the porosity is frequently specified in terms of the apparent density ρ.

In Europe the apparent density of porous metal bearings is generally specified as the "dry density," i.e.,

$$\rho\,(\text{dry}) = \frac{\text{Weight of the unimpregnated part}}{\text{Volume of the part}} \tag{2a}$$

In the U. S. A. the density is generally specified as the wet density, i.e.,

$$\rho\,(\text{wet}) = \frac{\text{Weight of part plus lubricant}}{\text{Volume of the part}} \tag{2b}$$

A tolerance of ±3 to 4% on the mean density can usually be achieved in production.

Oil Content

The oil content of a porous metal bearing is defined as

$$\text{Oil content} = \frac{\text{Volume of oil}}{\text{Volume of the part}} \tag{3a}$$

Table 1

Material	Equation for K in psi
90 : 10 tin bronze	$K = 14{,}340\rho - 66{,}100$
Soft iron	$K = 16{,}110\rho - 71{,}810$
95 : 5 iron copper	$K = 23{,}950\rho - 101{,}100$
90 : 10 iron copper	$K = 32{,}730\rho - 141{,}600$

However, a more informative figure is the degree of fullness or "impregnation factor," which is

$$\text{I.F.} = \frac{\text{Oil content}}{\text{Porosity}} \tag{3b}$$

In practice it is possible to impregnate 85 to 95% of the theoretical porosity. The apparent density and impregnation factor together check that the forming and the impregnation processes have been correctly carried out.

Degree of Sintering

The effectiveness of the sintering process determines the strength developed in the finished product. Although porous metal bearings are not generally called upon to act as unsupported structural members, certain minimum strength requirements are necessary. The large majority of porous metal bearings have a shape based on a hollow cylinder, and a popular test for them is the "radial crushing strength test" [4].

Radial Strength

Allowance for the dimensions of the component is made according to

$$\text{Minimum radial strength in lb} = KLH^2/M \tag{4}$$

where K is a strength factor in psi (see Table 1), L is the length of the cylinder in inches, H is the wall thickness in inches (T in B.S. 1131), and M is the mean diameter in inches $(D+H)$. This formula is only approximate.

Table 1 gives a list of accepted minimum strength values for K for four popular bearing alloys, where ρ is the dry density.

Equation (4) makes assumptions regarding the ductility of the material, the mode of fracture, and the position of the neutral axis. The error in these assumptions is not great with thin-walled, large-diameter bearings, but with thick-walled bearings the error becomes significant.

Tensile Strength

With porosities between 20 and 35%, the tensile strength of porous metal is roughly $\frac{1}{2}$ to $\frac{1}{4}$ of that of solid metal. This loss of strength arises from a combination of the following factors:

1. Reduced quantity of metal, due directly to the presence of porosity (this is the major effect).
2. The ratio of the cross-sectional area of the welds between particles to the overall cross-sectional area is less than the ratio of the volume of metal to the overall volume.
3. The notch effect of the porosity produces local high stress concentrations at the welds between particles.

These three factors contribute toward a lower strength than would be expected from a strictly linear relationship between porosity and strength.

Permeability

The size of the pores and the quantity of porosity determines the permeability of the porous metal. For further details references 5, 6, 8, and 14 should be studied, but it will suffice to quote here the law governing the flow of viscous fluids through typical porous bearing materials.

Darcy's law [7], which incidentally has not yet been fully derived from first principles, states that the flow rate/unit surface area, i.e., particle velocity w is

$$w = -\frac{\partial p}{\partial z}\cdot\frac{\Phi}{\eta} \tag{5}$$

where w is the velocity in the z direction, i.e., flow rate/unit area (Q/A), $\partial p/\partial z$ is the pressure gradient in the direction of flow, η is the absolute viscosity, and Φ is the permeability, which has units of area.

Departure from this empirical law occurs when the velocity is high (i.e., when the Reynolds numbers are greater than about 10), and when the pore size approaches the mean free path of the molecules of the fluid [8]. In all instances involving the flow of oil in porous metal bearings neither of these two exceptions operate, and Darcy's law can be applied with confidence.

The permeability is most conveniently measured using air as the test fluid. For a cylindrical bearing permeating from bore to O. D.,

$$\Phi = \left(\frac{Q}{\delta p}\cdot\frac{\eta}{L2\pi}\right)\log_e\left(1+\frac{2H}{D}\right) \tag{6}$$

where Q is the volume flow rate measured at the mean pressure and temperature (cc/sec), δp is the pressure drop (dynes/cm^2), η is the absolute viscosity (about 180×10^{-6} poise for air), H is the wall thickness, D is the bore diameter, L is the length (cm), and Φ is the permeability (cm^2).

The only standard unit of permeability is the darcy. This is defined in Europe as 10^{-8} cm^2. For porous metal bearings the darcy is inconveniently large, and to avoid confusion, it is most frequently expressed as 10^{-12} cm^2 units. Typical materials range from $\Phi = 100 \times 10^{-12}$ cm^2 to 2000×10^{-12} cm^2.

The relationship between porosity and permeability depends upon the tortuosity of the channels and the shape of the pores, but generally for a given porous material there is approximately a linear relationship between log permeability and log porosity [9, 14]. In porous metal bearings both these factors can be altered within certain limits.

Stress – Strain Characteristics

When nonporous metals deform plastically, the volume (or density) remains virtually constant. With porous metals, however, a considerable amount of plastic deformation can be accommodated by a reduction in volume, corresponding to a fall in porosity and a rise in density toward the value for the nonporous metal.

In the fully annealed condition the compressive stress – strain diagram for porous metal is a continuous curve as shown in Fig. 2. During manufacture, the coining operation produces a quantity of

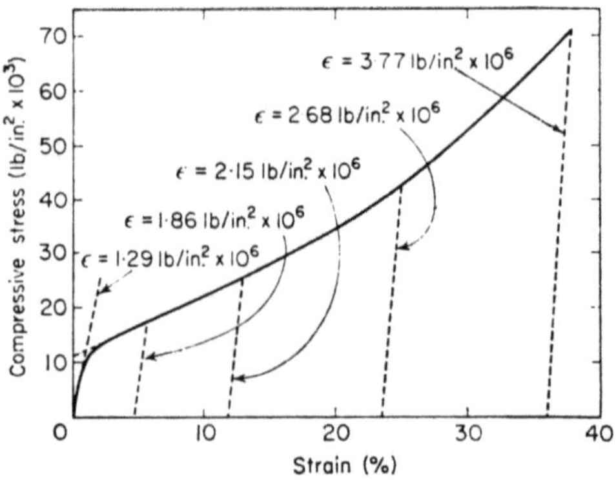

Fig. 2. Stress – strain curve for porous metal in compression, showing the change in modulus at relaxation with increasing strain.

permanent set and hence increases the elastic range. To produce further plastic deformation, a compressive stress greater than that used in the coining operation has to be used.

Other Physical Properties

The thermal and electrical conductivity of porous metal is less than that of nonporous metal. This arises from the presence of porosity per se, and from the limiting cross-sectional area of the welds between particles, as with the tensile strength. A similar reduction in magnetic permeability is observed with porous iron due to the air gap effect of the porosity. The coefficient of thermal expansion is, however, not affected by the porosity.

Porous Bearing Theory

The mathematical formulation of a solid shaft running in a porous housing has been given in reference 10. It follows directly from Reynolds' equation in three dimensions, which may be written

$$\frac{\partial}{\partial x}\left(h^3\frac{\partial p}{\partial x}\right)+\frac{\partial}{\partial y}\left(h^3\frac{\partial p}{\partial y}\right) = 6\eta\left[U\frac{dh}{dx}+2(w_h-w_0)\right] \tag{7}$$

where w_h and w_0 are the velocities of the top and bottom surfaces.

If either of these surfaces is permeable then w_0 and w_h are the velocities of the fluid passing through the porous walls. A glance at Fig. 3 showing how the continuity equation is derived makes this clear.

Now the velocity of fluids equals the volume of flow per unit area; hence Equation (5) above may be written

$$w = -\frac{\partial p}{\partial z}\cdot\frac{\Phi}{\eta} \tag{8}$$

The negative sign indicates that the direction of flow is in the direction of decreasing pressure. Hence if the top boundary (the shaft) is solid, then $w_h = 0$, and putting Equation (8) into Equation (7) gives

$$\frac{\partial}{\partial x}\left(h^3\frac{\partial p}{\partial x}\right) + \frac{\partial}{\partial y}\left(h^3\frac{\partial p}{\partial y}\right) = 6U\eta\frac{\mathrm{d}h}{\mathrm{d}x} + 12\Phi\left(\frac{\partial p}{\partial z}\right)_0 \tag{9}$$

where $(\partial p/\partial z)_0$ shows the outward pressure gradient at the oil/matrix interface.

In journal bearings $h = c(1 + \varepsilon\cos\theta)$ and $x = R\theta$, so the equation becomes, writing $(1 + \varepsilon\cos\theta) = h^*$ and $Ly^* = y$,

$$\frac{\partial}{\partial\theta}\left(h^{*3}\frac{\partial p}{\partial\theta}\right) + \frac{R^2\partial}{L^2\partial y^*}\left(h^{*3}\frac{\partial p}{\partial y^*}\right) = \frac{6U\eta R}{c^2}\frac{\mathrm{d}h^*}{\mathrm{d}\theta} + 12\frac{\Phi R^2}{c^3}\left(\frac{\partial p}{\partial z}\right)_0 \tag{10}$$

In order to nondimensionalize the equation completely, p^* is defined as $(c^2/6U\eta R)\times p$ and $Rz^* \equiv z$, so that

$$\frac{\partial}{\partial\theta}\left(h^{*3}\frac{\partial p^*}{\partial\theta}\right) + \frac{R^2\partial}{L^2\partial y^*}\left(h^{*3}\frac{\partial p^*}{\partial y^*}\right) = \frac{\mathrm{d}h^*}{\mathrm{d}\theta} + 12\frac{\Phi R}{c^3}\left(\frac{\partial p^*}{\partial z^*}\right)_0 \tag{11}$$

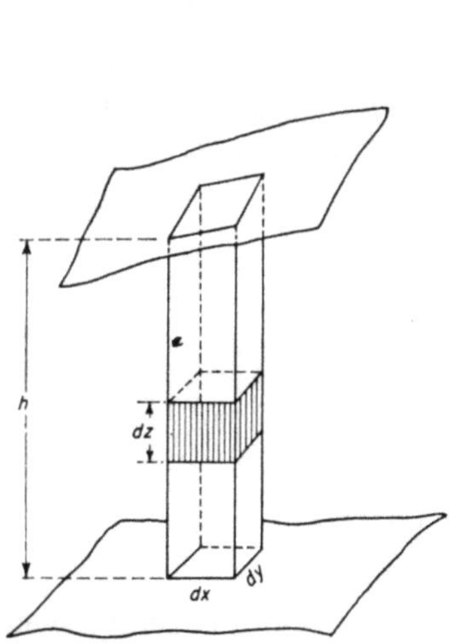

Fig. 3. Flow in a column.

In the matrix the continuity of flow equation is

$$\frac{\partial u}{\partial x}+\frac{\partial v}{\partial y}+\frac{\partial w}{\partial z} = 0$$

and the velocities, u, v, and w, are simply

$$u = -\frac{\partial p}{\partial x}\frac{\Phi}{\eta} \quad v = -\frac{\partial p}{\partial y}\frac{\Phi}{\eta} \quad w = -\frac{\partial p}{\partial z}\frac{\Phi}{\eta}$$

giving

$$-\frac{\Phi}{\eta}\times\nabla^2 p = 0 \tag{12}$$

where ∇^2 is the Laplace operator, $\partial^2/\partial x^2+\partial^2/\partial y^2+\partial^2/\partial z^2$. The solution of this Laplace equation in the matrix gives the value of $(\partial p/\partial z)_0$ at the interface.

It is possible to obtain a solution of Equations (11) and (12) for the infinitely short bearing, in the following way.

Neglect the first term of (11); this leads to a parabolic pressure distribution in the axial, y, direction.

The Laplace equation simplifies down to $\partial^2 p/\partial y^2+\partial^2 p/\partial z^2=0$ if $\partial^2 p/\partial x^2$ is neglected, and this two-term equation is satisfied by such a parabolic pressure distribution. Hence

$$\frac{\partial^2 p}{\partial z^2} = -\frac{\partial^2 p}{\partial y^2} \quad \text{and} \quad \frac{\partial p}{\partial z} = \left(C-\frac{\partial^2 p}{\partial y^2}\times z\right)$$

The pressure gradient is taken as zero on the outside boundary of the matrix, of radial thickness H, which is considered to be encased in an impermeable housing. Thus $\partial p/\partial z=0$ at $z=-H$ (remembering that z is positive in the direction of the oil film thickness), giving

$$\frac{\partial p}{\partial z} = -\frac{\partial^2 p}{\partial y^2}(H+z)$$

so at the interface, $z = 0$

$$\left(\frac{\partial p}{\partial z}\right)_0 = -H\left(\frac{\partial^2 p}{\partial y^2}\right) \tag{13}$$

and recalling that $Rz^* = z$ and $Ly^* = y$ and dividing the pressure by the requisite factor to render it nondimensional, the relation reads

$$\frac{1}{R}\left(\frac{\partial p^*}{\partial z^*}\right)_0 = -\frac{H}{L^2}\cdot\frac{\partial^2 p^*}{\partial y^{*2}}$$

which, as the $(\partial p/\partial x)$ term is neglected, can be put into (11) to give

$$\frac{R^2}{L^2}\cdot\frac{\partial}{\partial y^*}\left(h^{*3}\frac{\partial p^*}{\partial y^*}\right) = \frac{dh^*}{d\theta}+12\frac{\Phi R}{c^3}\left(-\frac{HR}{L^2}\frac{\partial^2 p^*}{\partial y^{*2}}\right)$$

Now h is not a function of y, so

$$\frac{\partial}{\partial y^*}\left(h^{*3}\frac{\partial p^*}{\partial y^*}\right) = h^{*3}\frac{\partial^2 p^*}{\partial y^{2*}}$$

Rearranging gives

$$\left(\frac{R^2}{L^2}h^{*3}+12\frac{\Phi H}{c^3}\cdot\frac{R^2}{L^2}\right)\frac{\partial^2 p^*}{\partial y^{*2}} = \frac{dh^*}{d\theta}$$

or

$$\frac{R^2}{L^2}\cdot\frac{\partial^2 p^*}{\partial y^{*2}} = \frac{dh^*/d\theta}{(h^{*3}+12\Phi H/c^3)} \tag{14}$$

Now $\Phi H/c^3$ is nondimensional, Φ having the dimensions of area, and is written as Ψ, to give

$$\frac{R^2}{L^2}\frac{\partial^2 p^*}{\partial y^{*2}} = \frac{dh^*/d\theta}{(h^{*3}+12\Psi)}$$

where $h^*=1+\varepsilon\cos\theta$ and $dh^*/d\theta=-\varepsilon\sin\theta$, whence

$$\frac{R^2}{L^2}\cdot\frac{\partial^2 p^*}{\partial y^{*2}} = -\frac{\varepsilon\sin\theta}{\{(1+\varepsilon\cos\theta)^3+12\Psi\}} \tag{15}$$

which on integrating twice, putting $p=0$ at $z=\pm L/2$ $(z^*=\pm\frac{1}{2})$, and $\partial p/\partial z=0$ at $z=z^*=0$, gives

$$\frac{R^2}{L^2}\times p^* = -\frac{\varepsilon\sin\theta}{2\{(1+\varepsilon\cos\theta)^3+12\Psi\}}\{(z^*)^2-\tfrac{1}{4}\} \tag{16}$$

This equation is interesting in that when the eccentricity ratio is unity the denominator is not zero at $\theta=\pi$, as it would be with a solid bearing. The pressure does not, therefore, tend to infinity, and so a finite load is carried.

The value of the parameter $\frac{\eta U}{W/L} \times \left(\frac{R}{c}\right)^2$, the Sommerfeld reciprocal, when $\varepsilon = 1$ is called the critical Sommerfeld reciprocal, $1/\Delta_c$. If $1/\Delta$ is increased beyond $1/\Delta_c$ then ε is just below unity, as the shaft is borne on an oil film. The value of $1/\Delta_c$ is a function of Ψ only on this simple theory. It has been computed out numerically in ref. 11.

The physical significance of Equation (16) is that at low speeds all the oil has leaked away into the matrix before the position of minimum film thickness (at $\theta = \pi$). The shaft has to rotate at a certain speed before it can carry sufficient oil round with it to permit leakage into the matrix while still having enough left to give a film separating the two surfaces at $\theta = \pi$.

Now for a full fluid film the coefficient of friction, μ, is given by

$$\mu \frac{R}{c} = \frac{2\pi}{(1-\varepsilon^2)^{1/2}\,\Delta}$$

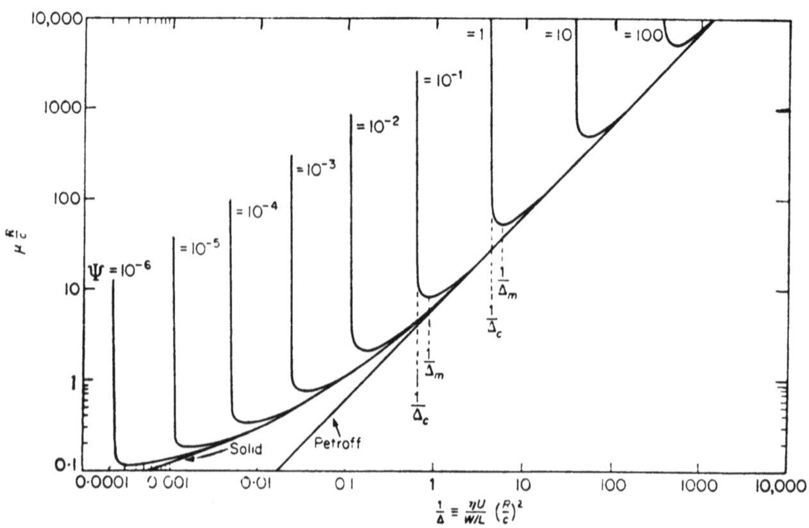

Fig. 4. Coefficient of friction against Sommerfeld reciprocal for range of ψ values, for narrow porous metal bearings, and showing $1/\Delta_c$ and $1/\Delta_m$.

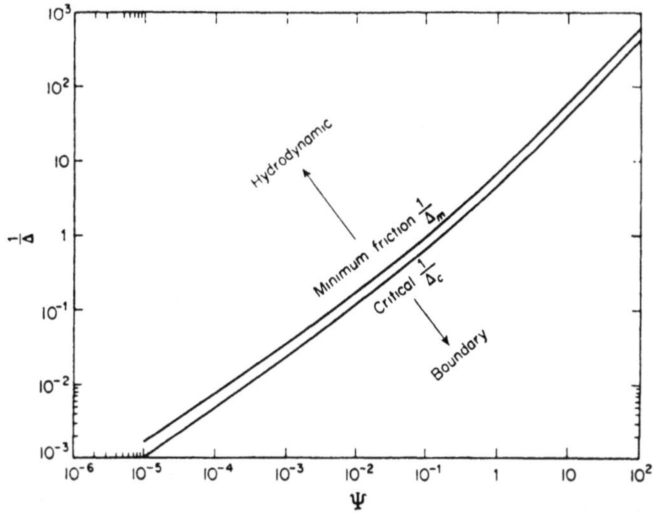

Fig. 5. Relationship between the critical Sommerfeld reciprocal, $1/\Delta_c$; Sommerfeld reciprocal for minimum friction, $1/\Delta_m$; and the porosity factor, Ψ, for narrow porous metal bearings.

Figure 4 gives $\mu R/c$ against $1/\Delta$ for a range of Ψ values. If the shaft is running at an eccentricity of unity when $1/\Delta = 1/\Delta_c$ the friction will be infinite, as $1-\varepsilon^2=0$. At some value of $1/\Delta$ just greater than $1/\Delta_c$ (with $\varepsilon < 1$) the friction is a minimum. The value of $1/\Delta$ to give minimum friction, $1/\Delta_m$, is the basis of a method of design.

While the full simultaneous solution of Equations (11) and (12) is not yet complete, a number of solutions are available for the infinitely long case, i.e., when the $\partial p/\partial y$ term is neglected with respect to $\partial p/\partial \theta$. The computations show that the parameter $\Phi H/c^3$ is the correct one to use providing $H < R/2$. Any increase of H over $R/2$ does not affect the operation of the bearing (all the pressure flow being near the inner surface) and so the value of Ψ is merely $\Phi R/2c^3$ whatever the value of H when it is larger than $R/2$.

In reference 11 the value of the critical $1/\Delta$ (i.e., $1/\Delta_c$) for various Ψ values is given. These are shown in Fig. 5 together with $1/\Delta_m$.

Bearing Performance

Basic Mechanism

The oil contained in the pores in the wall of a porous metal bearing is intended to last a finite time, sometimes for the life of the machine to which it is fitted, and sometimes for a specified period before replenishment is recommended. This oil wets the surface of the bearing and the shaft which, by rotating, sets up a circumferential pressure gradient. Oil is thus drawn from the unloaded region to form the usual pressure wedge, at which point some passes into the pores of the bearing and is recirculated to the unloaded region under the action of the induced pressure gradient.

The normal rise in temperature increases the volume of oil available, and when the rotation ceases the oil contained in the clearance is reabsorbed into the pores by capillarity, for future use. Under conditions which are unfavorable for the formation of a pressure wedge of sufficient magnitude to support the shaft (i.e., below the critical Sommerfeld reciprocal, $1/\Delta_c$, shown in Fig. 5 and discussed later) the pores act as pockets for the oil, and the usual condition of marginal lubrication is developed immediately as the shaft rotates and is readily maintained thereafter. It is possible that the land between pores produces a miniature hydrodynamic wedge, although the performance is then of course less than when conditions are favorable for full fluid lubrication.

Temperature

The oil in a porous metal bearing cannot carry away the frictional heat as it does in a pressure-fed bearing. Instead the heat is dissipated through the housing and along the shaft by conduction. The oil is therefore maintained at near the bearing running temperature for the entire working life of the bearing. As the quantity of oil available is comparatively small, the life of the bearing is dictated by the life of the oil at the operating temperature.

For this reason the design of porous metal bearings is directed toward a low running temperature. Apart from the adverse effect of high temperature on the oxidation resistance of the oil and the rate at which it may be lost from the pores, a low running temperature is desirable because of the effect of differential thermal expansion on the running clearance.

Clearance

The running clearance is usually determined from measurements made at room temperature, but the predicition of bearing performance involves the clearance existing at the actual operating temperature. Correction to the room temperature clearance has therefore to be made to compensate for the effects of both an ambient temperature different from room temperature and the temperature rise due to friction. The effect of these two temperatures will not necessarily be the same.

The materials of the shaft and the bearing (and the housing, if the bearing is force fitted) will frequently have different coefficients of thermal expansion. Where the ambient temperature is different from room temperature a simple correction can be made, but attention should be given to the loss of interference and the attendant possibility of a force-fitted bearing becoming loose in its housing. In the particular case of a rigid housing of lower coefficient than the porous bearing and an ambient temperature above room temperature, it should be remembered that some of the excess volume of porous bearing material produced in the expansion will generally be transmitted to the bore diameter of the bearing, to contribute toward a reduction in clearance. In the correction for the temperature rise due to friction, the influence of the housing on the expansion process is a function of both the thermal conductivity and the wall thickness of the bearing. No method is yet available for obtaining a reliable estimate of the net changes in clearance with temperature under all possible conditions.

The correction for changes in clearance is particularly important at high P. V. values and small clearance ratios, which are often used (sometimes ill-advisedly), for high-speed applications in order to minimize the out-of-balance forces. The influence of clearance on bearing performance is discussed elsewhere, but it should be noted that the porosity factor Ψ for porous bearings varies inversely as the cube of the radial clearance [12].

P. V. factor

It is clear that the power loss or rate at which work is done at the bearing surface is given by the product of the friction loss and the rubbing velocity:

$$\text{Power loss} = \text{Friction loss} \times \text{V.} = \text{rate of heat input} = W\mu\text{V} \qquad (17)$$

The output of heat by conduction is obtained from the product of the conductivity K, the effective area normal to the heat flow A_1, and the temperature difference δT. Thermal equilibrium is given by

$$WV\mu \propto KA_1\,\delta T$$

If the projected area of the bearing is A_2 we have

$$\delta T \propto \frac{\text{P.V.}\,\mu A_2}{KA_1} \tag{18}$$

For a rough estimate of the bore and length for a given load, W, and a given rotational velocity, N, Equation (18) leads to the well-known P. V. formula [Equation (19)]. Taking average values of μ, A_1/A_2, and K, the safe running temperature can be expressed in terms of the P. V. factor, where P is the bearing pressure in psi and V is the surface velocity of the shaft in ft/min.

The average maximum value generally quoted is

$$\frac{\delta TKA_1}{\mu A_2} = \text{P.V.} = 50{,}000 \text{ psi } \times \text{ft/min} \tag{19}$$

However, the limiting range of P. V. is probably from 10,000 to 250,000 psi × ft/min.

The application of Equation (18) has, of course, other limiting conditions at extremes of P. and V. For example, the limiting pressure is governed by the fatigue and elastic limits of the porous metal, and the limiting velocity is governed by the amount of out-of-balance loading.

Within these conditions, the choice of oil viscosity, length-to-diameter ratio, running clearance, permeability, and wall thickness can be made to give the lowest value of δT in Equation (18), and thus a design for the lowest running temperature and longest life [12].

Amount of Underdesign

It is shown in reference 10 that operation below the critical value of $1/\Delta$ causes pore closure in the loaded region of the bearing surface, and that this results in an irreversible reduction in permeability Φ.

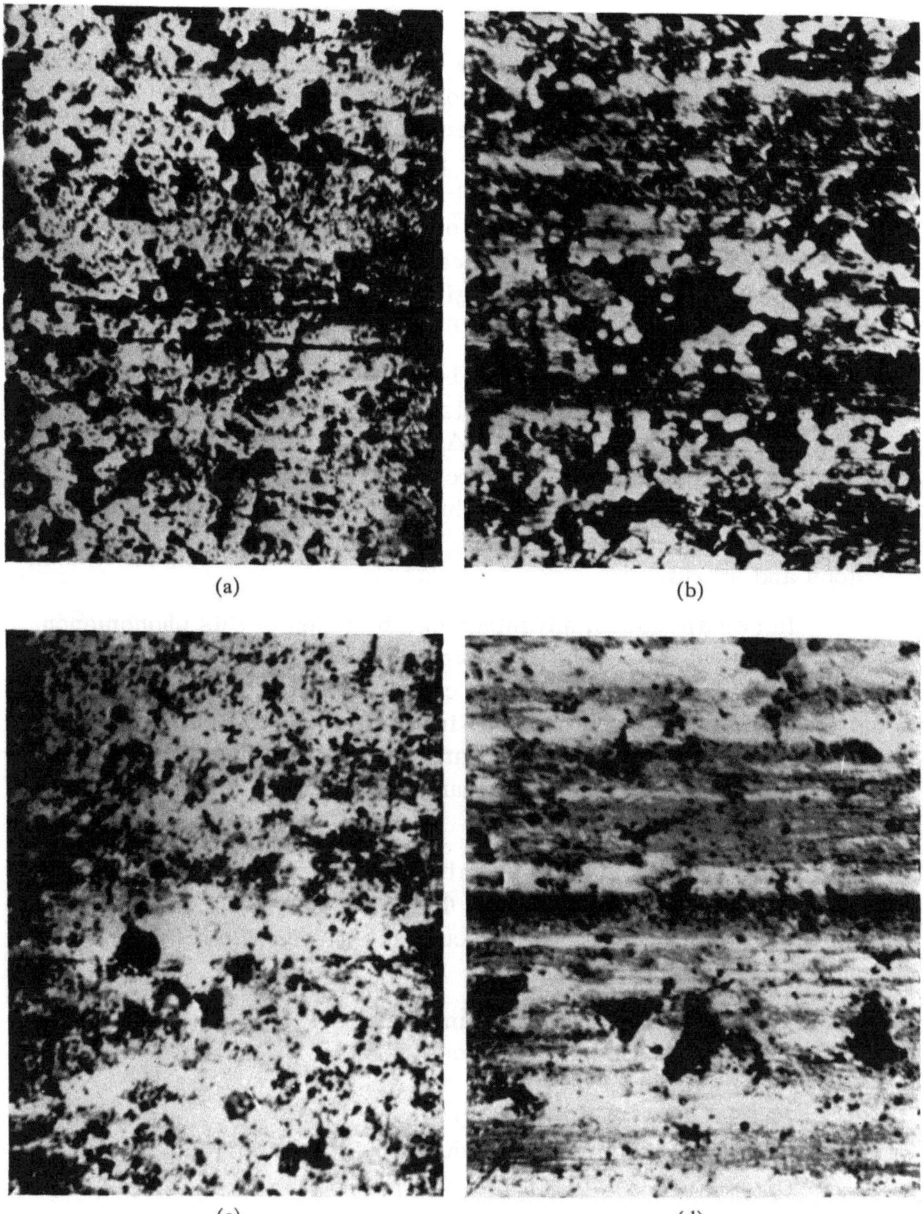

Fig. 6. Examples of pore closure in the loaded area, showing the reduction in permeability produced by deliberate running under boundary conditions: (a) $\Delta = 0.8$, $\psi = 5.8 \times 10^{-3}$; (b) $\Delta = 4.1$, $\psi = 2.9 \times 10^{-3}$; (c) $\Delta = 31.5$, $\psi = 8.35 \times 10^{-4}$; (d) $\Delta = 43.5$, $\psi = 1.4 \times 10^{-4}$.

When the integrated oil film pressure is not greater than the applied bearing pressure, the shaft will not "float" but will rub on the bearing metal with a force equal to the difference between these two forces. The effect of this gentle rubbing in the presence of a very thin boundary film of oil is to cause plastic flow of the bearing surface metal and a reduction of the size of the surface pores. This usually occurs without measurable wear, and produces a fall both in permeability and in the amount of oil passing into the pores in that region. The process can be followed from the examples given in Fig. 6.

As the permeability of the bearing in the loaded region is reduced, the porosity factor is also reduced and approaches the value of Ψ, corresponding to $1/\Delta_c$, at a decreasing rate. When criticality is reached the two forces balance and the shaft and bearing become separated by a fluid film of oil. The process of pore closure then ceases, except, of course, each time the shaft stops and starts.

In certain cases, advantage can be taken of this phenomenon and a value of Ψ chosen for the initial running conditions such that a calculated amount of pore closure in the loaded region is allowed to occur during running-in. This underdesigning results in a lower ultimate value of μ, and is achieved by designing for a value of Ψ to give $1/\Delta$ less than $1/\Delta_c$, the critical Sommerfeld reciprocal.

There is undoubtedly a limit to the amount of pore closure (and corresponding reduction in Φ and hence Ψ) which can be tolerated during running-in without causing one of the following undesirable effects:

(a) High initial running temperature
(b) High initial oil oxidation
(c) High initial oil loss
(d) Seizure from differential expansion
(e) Debris from wear products
(f) Seizure from pick-up.

With typical porous bronze bearings running at R/c about 600 and P. V. of 50,000 psi × ft/min, an underdesign of about 100-fold can usually be tolerated in practice.

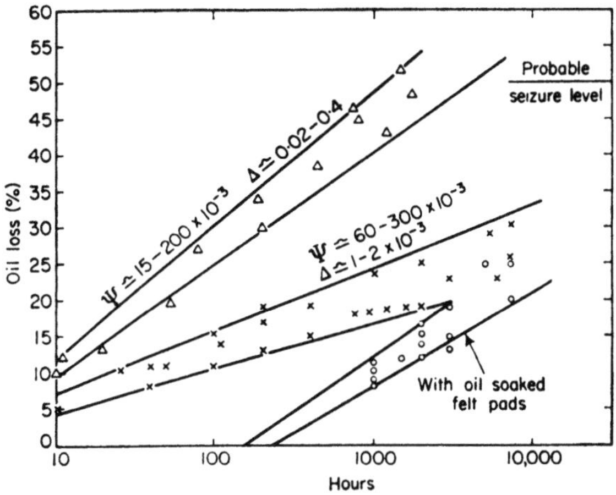

Fig. 7. Tests of oil loss as a function of time.

Quality of the Oil

The viscosity of the oil is chosen to give the required value of $1/\Delta$. The next important characteristic is oxidation stability. With porous metal bearings this requires special attention because the oil is maintained at very near the running temperature, porous metal has a lower thermal conductivity, the oil is in contact with a large area of metal capable of catalytic oxidation, and air, present in the unfilled pores, circulates with the oil.

High oxidation stability is achieved by refining and antioxidant additions, at the expense of reduced oiliness. The compromise between these two factors has to be made on the basis of the total thermal duty and the amount of running in the boundary friction region before fluid lubrication is established. For example, in an application involving continuous operation with a high running temperature, an oil having maximum oxidation stability would be chosen; on the other hand, in an application involving intermittent running or reciprocating motion and normal running temperature, an oil with "oiliness" additives would be chosen.

Attention to the pour point is necessary with applications likely to give cold starting troubles, and extreme pressure additives and detergents which operate by surface activity have to be carefully chosen in view of the very large available surface within the pores of the metal matrix.

Life

The life of a porous metal bearing is determined principally by the life of the oil, measured in terms of either quantity or chemical stability or both. The råte of oil loss is approximately exponential [13], as would be expected from a study of the surface tension forces produced within a range of pore sizes. Not all the oil content is consumed in the life of the bearing, and from practical tests it appears that not more than half of the oil can be lost without a significant fall in performance, although the limit undoubtedly depends upon the severity of the duty and the design details. Typical results are shown in Fig. 7.

Failure can occur by means other than oxidation or oil loss. If the position of closest approach of the shaft and bearing rotates with respect to the bearing, there is a danger that the pores at the entire surface of the bearing may become closed, so that circulation of oil to the rubbing surfaces is eventually prevented. This can arise with a rotating bearing and a unidirectional load, or a rotating load and a stationary bearing.

Whether ultimate failure is manifest as seizure or as wear depends upon the power of the driving motor and whether an increase in temperature produces an increase or decrease in running clearance.

Supplementary Lubrication

The total volume of oil available for lubrication at the bearing assembly can be increased by the use of some absorbent material in contact with the outside of the porous metal. Wool felt is frequently used for the absorbent material. If the porous metal has a higher entry capillary pressure [14] than the felt, oil will flow only toward the porous metal. The porous metal is thereby kept fully charged with oil, and the rate of oil loss is temporarily increased until the extra oil contained in the felt has been consumed by the bearing. This flush of extra lubricant is, of course, useful during the running-in of the bearing.

The use of a felt reservoir also facilitates recharging from a simple hole. The felt readily absorbs the replenishing oil and transfer to the fine pores of the bearing can then take place over an extended period.

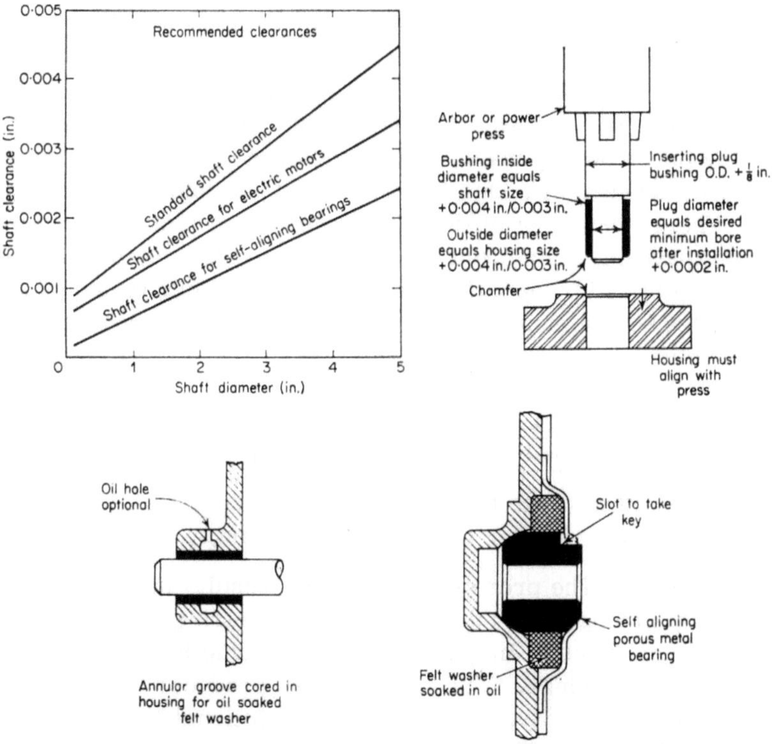

Fig. 8. Bearing clearances and fitting.

Fitting

Porous metal bearings are secured into the housing either by a force fit, or as an insert in the plastic molding or die casting of the housing, or (in a self-aligning assembly) by spring-loaded pressure plates on the spherical outer surface of the self-aligning bearing, as shown in Fig. 8.

When a cylindrical bearing is force fitted into the housing a proportion of the interference fit is transmitted to the bore. The ratio of the interference to the bore closure depends upon the rigidity of the housing, the strength and density of the bearing material, and the wall thickness of the bearing. Allowance has to be made for this in dimensioning the bearing to give the correct running clearance.

Porous metal bearings cannot be machined or reamed without closure of the porous surface. Hence the size of the bearing is designed so that the bore is the correct dimension when fitted. Nonworking surfaces, such as the outside diameter and length, can be machined without detriment to the bearing performance, provided always that the swarf is effectively removed and the oil lost from the pores during machining is replaced by oil impregnation. Alignment between bearings is achieved by alignment of the housings. Where machining of the fitted bearing bore cannot be avoided for some reason, the amount of pore closure and loss of efficiency can be minimized by adopting the special machining techniques recommended by the manufacturers for the grade of porous metal alloy involved.

Limitations and Advantages of Porous Metal Bearings

The limitations to the application of porous metal bearings can be summarized under the following headings.

Porosity. The presence of porosity results in a loss of mechanical strength, so that porous metal bearings are less useful under conditions of impact loading or fatigue loading than nonporous metal bearings. The porosity also produces a reduction in thermal conductivity, so that the total frictional losses which can be safely dealt with are also less. An increase in the size of the bearing to compensate for these factors is sometimes possible. The porosity also results in a loss of pressure in the oil film. The magnitude of this loss can be determined and allowed for in the design.

Size. The overall size of porous metal bearings is limited at one end of the scale by the capacity of available automatic presses, and by the size of the particles of metal powder at the other end. The size range is roughly from about $\frac{1}{32}$ of an inch to several inches, with the majority of parts in the range $\frac{1}{8}$ of an inch to 3 inches diameter.

Shape. The shape of porous metal bearings is limited to those which can be formed by axial compression, an inherent limitation of the powder metallurgy process. In order that the porosity shall be reasonably uniform along the axis of compression, there is a limit to the length-to-wall-thickness ratio that can be produced successfully [15, 16].

Composition. The composition of the porous metal alloy is limited to alloys amenable to the powder metallurgy technique.

Thus reduction of the metal oxide at temperatures below the melting point must be feasible with conventional reducing atmospheres, such as CO, H_2, etc. The metal must also be capable of production in powder form sufficiently ductile to permit pressing. Hence Cu, Fe, Ni, W, Ag, Pb, Sn, etc. are suitable, and Al, Cr, Mg, Mn, Si, V, etc. are difficult metals to fabricate by this technique. Because the starting materials are powders, nonmetallic powders and solid lubricants which can survive the sintering operation, such as graphite, MoS_2, etc., can be readily and homogeneously incorporated into the composition.

The advantage of porous metal bearings can be summarized as follows.

A self-lubricating bearing overcomes the need for oil pipes, pumps, etc., and simplifies the problems of machine design. Oil drip and the possibility of spoiling manufactured articles is reduced to a minimum. Bearings can be accommodated in positions which are not readily accessible, and the need for frequent maintenance is overcome. They are easily fitted, and as a result of mass production are cheap, often cheaper than bearings machined from castings.

Selecting a Bearing Material

Bearing problems arise whenever the mechanical engineer has to transmit power from one point to another. Similar problems face the electrical engineer in the transmission of electrical power, and the insulators he uses have to be chosen with the same care as the bearings used by the mechanical engineer. Their functions are very similar and the outcome of their failure in service no less serious.

In the same way that there is no single best electrical insulating material, there is no single best bearing material. Each has its own field of application. The foregoing notes have attempted to define the field occupied by porous metal bearings by indicating their limitations. However, in common with many other successful developments, the "theory" has followed (and is still following) the trail of "practice." It will be clear that all is not yet known, but it is hoped that in setting out the present state of our knowldege, the bearing designer is better equipped to apply porous metal bearings to his problems. Even so it has only been possible to present the general case. Each manufacturer has developed a number of particular porous metal alloys and it is strongly recommended that the designer should be guided by the advice of the manufacturer in these particular cases.

References

1. Geotzel, C. G., Treatise on Powder Metallurgy, Interscience Publishers, New York, 1950.
2. Powder Metallurgy Joint Group of Iron and Steel Institute and Institute of Metals, London Biannual publications, 1957 onwards.
3. Jones, W. D., Fundamental Principles of Powder Metallurgy, Edward Arnold, London.
4. B. S. S. 1131 Part 5, 1955. British Standards Institution, London.
5. Muskat, M., The Flow of Homogeneous Fluids through Porous Media, J. W. Edwards Inc., Michigan, 1946.
6 Barrer, R. M., Diffusion in and through Solids, University Press, Cambridge, 1951.
7. Darcy, H P C., The Water Supply of the Town of Dijon (ed. Victor Delmont), Paris (Libraire des Corps Imperiaux du ponts et chaussées et des mines), p. 590.
8. Morgan, V. T., "Filter elements by powder metallurgy," Symposium on Powder Metallurgy, Iron and Steel Institute Special Report, Vol. 58, 1954.
9. Grootenhuis, P. and Leadbeater, C. J , Discussion, Symposium on Powder Metallurgy, Iron and Steel Institute Special Report, Vol. 58, pp. 361-363, 367-369, 1954.
10. Morgan, V. T. and Cameron, A., "Mechanism of lubrication in porous metal bearings," Conference on Lubrication and Wear, Institution of Mechanical Engineers (1957), London, pp. 151-157.
11. Cameron, A., Morgan, V. T., and Stainsby, A E , "Critical conditions for hydrodynamic lubrication of porous metal bearings," Proc. Inst. Mech. Eng., Vol. 176, pp. 761-770, 1962.
12. Morgan, V. T., Hydrodynamic Porous Metal Bearings, A. S. L. E., Chicago, Illinois, 1964.
13. Morgan, V. T. "Study of the design criteria for porous metal bearings," Conference on Lubrication and Wear, Institution of Mechanical Engineers (1957), London, pp. 405-408.
14. Eisenklam, P., "Porous masses," Section 9. "Chemical Engineering Practice," Vol. 2 , Solid State (ed. H. W. Cremer and T. Davies), Butterworths Scientific Publications, London.
15. Victor M. T. and Sorg, C. A., Metals and Alloys, Vol. 19, No. 3, p. 584, 1944.
16. Morgan, V. T., "The effect of porosity on some of the physical properties of powder metallurgy components," Powder Metallurgy, Inst. Metals, London, Vol. 12, 1963.

Chapter 16

Lubrication of a Porous Bearing—Stokes' Solution

D. D. Joseph

Department of Aeronautics
and Engineering Mechanics
University of Minnesota
Minneapolis, Minnesota

L. N. Tao

Department of Mechanics
Illinois Institute of Technology
Chicago, Illinois

Introduction

The problem of finding the pressure distribution and force resultants created by the rotation of an impermeable journal in a permeable bearing is of practical engineering importance. Many bearings of this type are used in the construction of machine elements. In this paper, we treat an idealized version of this problem with the aim of isolating the physical effects of the bearing flow,

There are several approaches to this and related problems which take the one-dimensional equations of Reynolds as a starting point. In the case of a viscous fluid which is entrained in the gap between rotating porous rollers, it has been found – by G. I. Taylor and J. C. Miller [1] for zero clearance and by L. N. Tao and D. D. Joseph [2] for finite clearance – that the loss of fluid through the porous surface reduces the magnitude and shifts the position of the cross section of maximum suction. This suggests that the introduction of porous materials may be useful in applications where high suction is undesirable, e.g., where cavitation is a problem. An essential defect of the approach used by [1] and [2] is that the flow through the permeable surface is assumed proportional to the pressure. According to Darcy's law, the through-flow should be locally proportional to the transverse gradient of pressure. It is by means of the transverse gradients that the fluid in the channel between journal and bearing is sucked from or forced into the channel; since the one-dimensional treatment cannot accommodate transverse pressure gradients, it is not capable of describing the through-flow in any but ad hoc fashion.

A treatment of the related problem of the lubrication of a porous bearing which introduces the through-flow in a more natural way is due to V. Morgan and A. Cameron [3] and to A. Cameron, V. Morgan, and A. Stainsby [4].

Though these authors start from the Reynolds' equation, they do require that the through-flow at the porous surface be related to the pressure in the bearing by Darcy's law. Darcy's law implies, as a consequence of the conservation of mass, that the pressure in the porous bearing be harmonic. This condition is also approximately satisfied by Cameron et al. The strength of their treatment is in the attempt to treat the flow as coupled. Of course, the transverse variation of pressure in the gap is necessarily neglected by virtue of the averaging implied by the Reynolds' equation. But, in addition to this, the solution has a very restricted range of applicability and is valid only for very narrow bearings in which axial variations of pressure dominate circumferential variations. For longer bearings (these are typical), circumferential variations are dominant. A solution of the modified Reynolds' equation proposed by Morgan and Cameron [3] which accounts for circumferential variations of pressure and neglects the axial variations has been constructed by Shir and Joseph [5].

In this paper, we will consider the porous bearing problem from a quite different point of view. Here, we seek to find the variations of pressure and the force resultants in the context of a coupled boundary-value problem in which different flow regimes are matched at common porous boundaries. For the bearing problem, we require that the flow in the bearing satisfy Darcy's law, and that the flow in the lubricating film satisfy the low Reynolds number approximations of Stokes. For the clearances which prevail in typical lubrication problems, the Reynolds numbers are typically small enough so that Stokes' approximations are valid. The solutions in the two regimes, the bearing and the liquid film, are to be matched at the bearing surface by the requirement of continuity as applied to the normal component of stress and velocity.

The idea of using Stokes' equations for examination of journal-bearing lubrications is not new. In fact, G. Wannier [6] has shown, for the nonporous bearing, how the Reynolds – Sommerfeld solution of Reynolds' equation may be recovered from a Stokes solution taken in the limit of small clearance. Applied to the permeable bearing, this method has the obvious advantages of a fully two-dimensional treatment. In addition, it is possible to obtain a Reynolds – Sommerfeld solution as a limiting case of the Stokes solution. One may in this way obtain an appropriate Reynolds – Sommerfeld solution without a corresponding Reynolds equation. It is this procedure which we adopt here.

For mathematical simplicity, we shall simulate a thick porous bearing with a fluid-saturated porous space with an internal hole in which a journal rotates. The harmonic pressure dies exponentially with distance from the internal hole, and the configuration can be shown to simulate the thick bearing with increasingly good accuracy as the radius ratio increases from a value of about three-halves [5]. Viewed in another way, the finite-bearing solution of the Reynolds equation used in [5] has greater credibility by virtue of the fact that it reduces properly to the Stokes solution as the radius ratio tends to infinity.

It is assumed that conditions of fully hydrodynamic lubrication prevail, that the permeable bearing is homogeneous and isotropic, and that the flow within it satisfies Darcy's law

$$\mathbf{Q} = -\frac{k}{\mu} \nabla P \tag{1}$$

where k, μ, and P are the permeability, dynamic viscosity, and pressure, respectively. The vector $\mathbf{Q}$ is a volume-flow rate per unit cross-sectional area. As such, it represents the filter velocity rather than the true velocity of the fluid in the pores. In the bearing

$$\nabla^2 P = 0 \tag{2}$$

as follows from mass conservation, $\nabla \cdot \mathbf{Q} = 0$.

One further assumption specifying some condition on the tangential component of velocity at the exterior of the bearing surface is required. Why this is so is made clear from the following considerations.

Consider a general flow in a layered medium satisfying Darcy's law in each of the separate layers. In these problems, which involve flow between materials of different permeability, boundary pressures and velocities are unknown and the boundary conditions are supplied by the requirement that both normal stress and the normal component of velocity are continuous across surfaces separating regions of different permeability. As in the hydrodynamics of ideal fluids, no requirements are imposed on the tangential components of velocity. This condition is consistent with the nature of the Darcy law, which postulates that changes in the empirically equivalent velocity $\mathbf{Q}$ do not depend on the mechanisms of shear. Obviously, the governing equation (2) is not of sufficiently high order to accommodate conditions on the tangential component of velocity, and a discontinuity of this component of velocity across the surface separating regions of different permeability is to be expected.

When the porous boundary separates a region of permeable material from a region where such material is absent, it is necessary, if the governing differential system is not be underdeterminate, to specify the behavior of the viscous liquid immediately outside the porous region.

In the case of a viscous flow, the order of the equations governing the external motion is raised by two, although the equations governing the internal motion are unchanged. It is therefore necessary to impose some requirement on the tangential component of velocity at the interface. When much of what would nominally be an impermeable surface is really hole entrance, the appropriate boundary condition is not easy to anticipate. However, when the permeability is small relative to some typical gross area of the porous body, it would seem valid to retain the no-slip condition relative to the wall. Since the permeability is slight in most natural materials, this condition is not severely restrictive.

In our present effort, we make the low Reynolds number (Stokes) approximations to the viscous equations. These approximations are particularly convenient in the coupled problem. Not only are the governing equations linear but both the external and internal pressure fields are harmonic, a fact which considerably lightens the labor involved in satisfying the matching conditions. For Stokes' classical problem (streaming of a viscous liquid past a porous sphere), the foregoing formulation is self-consistent, and a uniform approximation to the total velocity field can be obtained [7, 8]. A more difficult matching problem is involved in an attempt to assess the effects of the ground flow induced by the motion of a cylinder near the ground [9]. This latter problem is very closely related to the lubrication problem considered in this paper, and the physical effects of the ground flow model those of the bearing flow.

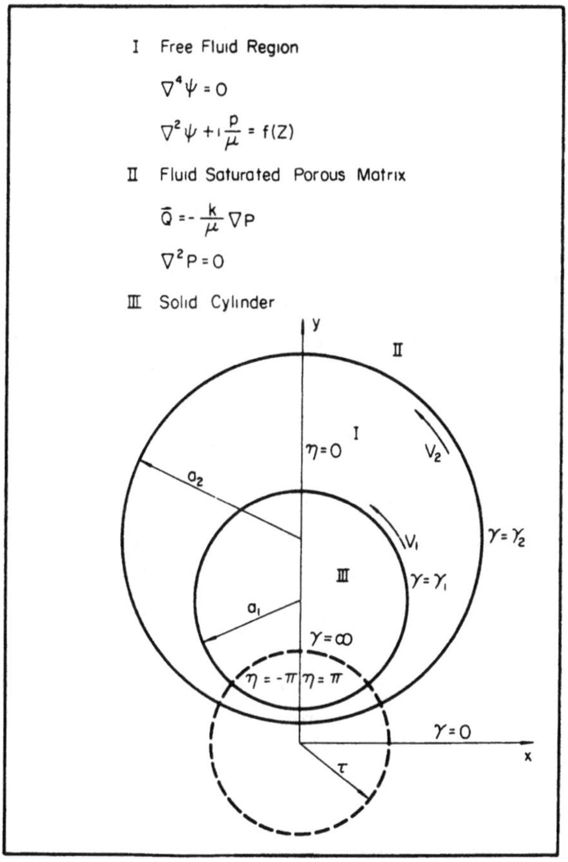

Fig. 1. Geometry of flow between nonconcentric cylinders.

The foregoing formulation is not restricted to a linear representation (Stokes' approximation) of the external flow, and at least one example of an external flow governed by the nonlinear Navier – Stokes equations (rotation of a porous disk in a viscous liquid) has been constructed [10].

The lubrication problem considered here may be represented geometrically by nonconcentric circles of a bipolar family (Fig. 1). For these, there are available classical results which apply in Stokes' regime (Re → ∞) when all boundary surfaces are impermeable. Although the general Stokes solution for the permeable case cannot be obtained in closed form, a very accurate compact approximation, equation (41), which is valid in the limit of small clearance can be derived. Moreover, in the practical range of permeabilities, this solution has almost the same circumferential variation as the known impermeable solution to which it reduces for vanishing permeability.

The solutions show that the principal physical effects introduced by the bearing flow are a diminution of pressure differences which would develop if journal and bearing were impermeable. The positions of maximum and minimum pressure are shifted away from the minimum gap. Total forces are altered not only by diminution of the pressure resultant but also by the relative insensitivity of the Couette-induced shear resultant to changes in pressure. This shift in the relative contributions of the pressure and shear resultants, which also operates on the induced ground-flow problem [9], can have an interesting effect on the load capacity of the bearing. For impermeable bearings, it is usual to identify the load capacity with the bearing capacity. This procedure, strictly speaking, is not valid, because there is a net contribution due to shear. However, the shear resultant is smaller than the pressure resultant by a factor proportional to the clearance and ordinarily can be neglected [11]. The shear-stress resultant is given by

$$X_s = 4\pi\mu a_1 V\{2 - (4 - \epsilon^2)/(2 + \epsilon^2)\alpha\}/e$$

where e is the eccentricity, $\epsilon = e/c$ is the eccentricity ratio, and $c = a_2 - a_1$ is the clearance. It may be readily verified that X_8 changes sign at $\epsilon = 0.721$ and is negative for larger values of ϵ. When the bearing is permeable, the pressure resultant may be greatly reduced. The shear-stress resultant varies slowly in magnitude and turns negative for values of the eccentric ratio which get smaller as k is increased (Fig. 2). With the pressure greatly

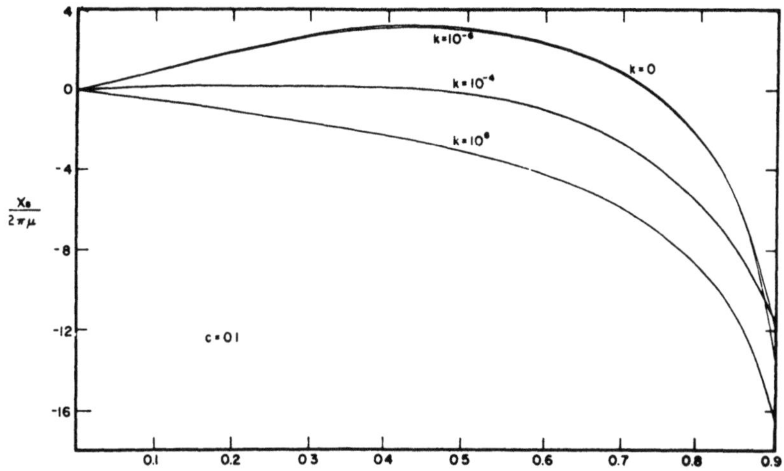

Fig. 2. Variation of shear-stress resultant with eccentricity and permeability. Shows graphs of equation (25) [or (38)]. For sufficiently large eccentricities (ϵ), this resultant opposes the sense of pressure resultant – much more so as permeability (k) is increased. For large ϵ and k, this resultant and pressure resultant are comparable.

reduced, the shear-stress resultant may become dominant and, for large eccentric ratios, the bearing capacity can in principle vanish and even change sign. It follows that a large load-carrying service is not to be expected from a porous bearing.

Coupled Flow Between Nonconcentric Cylinders

We consider the flow between a solid cylinder (III) and a cylindrical cavity in an infinite porous matrix (II). The low Reynolds number approximations of Stokes govern region I and the equations of Darcy apply in II. The effect of a rotating cylinder near a moving porous ground is obtained when $\gamma_2 = 0$. The cylinders are represented by bipolar circles (Fig. 1).

We introduce Lagrange's stream function for the fluid in region I. In the Stokes approximation, the pressure is harmonic and conjugate to the Laplacian of the stream function. Thus

$$\nabla^2\psi + ip/\mu = f(z) \tag{3}$$

is an analytic function of the complex variable $z = x + iy$. The x and y components of velocity are $-\partial\psi/\partial y$ and $\partial\psi/\partial x$, respectively.

We do not consider the rotation of the porous space, so that $V_2 = 0$ when $\gamma_2 \neq 0$.

The governing equations are

$$\nabla^4\psi = 0 \qquad \text{(in I)} \qquad (4)$$

$$\nabla^2\psi + ip/\mu = f(z)$$

$$\mathbf{Q} = -\frac{k}{\mu}\nabla P \qquad \text{(in II)} \qquad (5)$$

$$\nabla^2 P = 0^* \qquad (6)$$

The boundary conditions are as follows:

1. The pressure is a single-valued function of the bipolar circular variable η, where

$$\zeta = \gamma + i\eta = \log\frac{z + i\tau}{z - i\tau} \qquad (7a)$$

and τ is one-half the distance between pole points.

2. Pressure is continuous across porous boundaries†

$$P(\gamma_2, \eta) = p(\gamma_2, \eta) \qquad (7b)$$

3. The normal component of velocity vanishes at the solid boundary

$$q_\gamma(\gamma_1, \eta) = -\frac{1}{h}\frac{\partial\psi}{\partial\eta}(\gamma_1, \eta) = 0 \qquad (8)$$

and is continuous across the porous boundary

$$hq_\gamma(\gamma_2, \eta) = -\frac{\partial\psi}{\partial\eta}(\gamma_2, \eta) = -\frac{k}{\mu}\frac{\partial P}{\partial\gamma}(\gamma_2, \eta) \qquad (9)$$

where $h = \tau/(\cosh\gamma - \cos\eta)$ is a metric coefficient.

4. The fluid does not slip at the porous boundary

$$q_\eta(\gamma_1, \eta) = \frac{1}{h}\frac{\partial\psi}{\partial\gamma}(\gamma_1, \eta) = -V_1 \qquad (10)$$

* Lower-case letters (q, p) give velocity and pressure in I. Capital letters (Q, P) give these quantities in II.

† The difference between the pressure and normal stress in the exterior fluid is proportional to the Darcy coefficient (k) times the curvature of the boundary. We neglect this small difference for the bearing problem.

$$q_\eta(\gamma_2, \eta) = \frac{1}{h}\frac{\partial\psi}{\partial\gamma}(\gamma_2, \eta) = -V_2 \tag{11}$$

General solutions of equations (3)-(11) are developed in [9]. In this section, we list only those results which are used in subsequent deductions. First define

$$C_{N\gamma} = \cosh N\gamma, \qquad S_{N\gamma} = \sinh N\gamma$$

$$\begin{aligned} \Phi_0(\gamma) &= \varphi_{01}C_\gamma + \varphi_{02}S_\gamma - \varphi_{12}\gamma C_\gamma + \varphi_{04}\gamma S_\gamma \\ \Phi_1(\gamma) &= \varphi_{11} + \varphi_{12}\gamma + \varphi_{13}C_{2\gamma} + \varphi_{14}S_{2\gamma} \end{aligned} \tag{12}$$

$$\Phi_N(\gamma) = \varphi_{N1}C_{N-1,\gamma} + \varphi_{N2}S_{N-1,\gamma} + \varphi_{N3}C_{N+1,\gamma} + \varphi_{N4}S_{N+1,\gamma}$$

The stream function for the motion of the fluid in the cavity is given by

$$\psi = \frac{\tau}{C_\gamma - \cos\eta}\sum_{N=0}\Phi_N(\gamma)\cos N\eta \tag{13}$$

and the pressure by

$$\begin{aligned} \frac{p\tau}{2\mu} = -\varphi_{04}S_\gamma\sin\eta + \sum_{N=1}\{(1-N)(\varphi_{N2} + \varphi_{N+1,4})C_{N\gamma} \\ + (N+1)(\varphi_{N4} + \varphi_{N+1,2})C_{N\gamma} + (1-N)(\varphi_{N1} + \varphi_{N-1,3})S_{N\gamma} \\ + (N+1)(\varphi_{N3} + \varphi_{N+1,1})S_{N\gamma}\}\sin N\eta \end{aligned} \tag{14}$$

The harmonic pressure P in the fluid saturated porous space is given by *

$$\frac{P\tau}{2\mu} = \sum_{N=1} A_N^{(2)}e^{N(\gamma-\gamma_2)}\sin N\eta \tag{15}$$

The cavity and porous matrix solutions are related by boundary conditions which lead ultimately to ($i = 1, 2$)

$$A_1^{(i)} = -\varphi_{04}S_\gamma + 2(\varphi_{14} + \varphi_{22})C_{\gamma_i} + 2(\varphi_{13} + \varphi_{21})S_\gamma \tag{16}$$

$$\begin{aligned} A_N^{(i)} = \{(1-N)(\varphi_{N2} + \varphi_{N-1,4}) + (N+1)(\varphi_{N4} + \varphi_{N+1,2})\}C_{N\gamma_i} \\ + \{(1-N)(\varphi_{N1} + \varphi_{N-1,3}) + (N+1)(\varphi_{N3} + \varphi_{N+1,1})\}S_{N\gamma} \end{aligned} \tag{17}$$

$$E_1^{(i)} = \varphi_{12} + 2\varphi_{13}S_{2\gamma_i} + 2\varphi_{14}C_{2\gamma} = (-1)^{i-1}2K_iA_1^{(i)}S_\gamma \tag{18}$$

* This pressure is finite at $\gamma = -\infty$, which is a pole point in the interior of the porous material. In [9], equation (15) should be replaced with (15) above. The inappropriate (15) of [9] is a miscopy and does not change any other results in that reference.

$$E_2^{(i)} = \varphi_{02} + \varphi_{04}(\gamma_i + S_{\gamma_i}C_{\gamma_i}) - C_{\gamma_i}^2\varphi_{12}$$
$$= -C_{\gamma_i}V_i + (-1)^{i-1}K_iA_1^{(i)}S_{\gamma_i} \tag{19}$$

$$E_3^{(i)} = \varphi_{00} - \varphi_{04}S_{\gamma_i}^2 + \varphi_{12}S_{\gamma_i}C_{\gamma_i} + \varphi_{13}C_{2\gamma_i} + \varphi_{14}S_{2\gamma}$$
$$= S_{\gamma_i}V_i - (-1)^{i-1}K_i(A_2^{(i)} - C_{\gamma_i}A_1^{(i)}) \tag{20}$$

$$E_{4N}^{(i)} = \varphi_{N1}C_{N-1,\gamma_i} + \varphi_{N2}S_{N-1,\gamma_i} + \varphi_{N3}C_{N+1,\gamma_i} + \varphi_{4N}S_{N+1,\gamma}$$
$$= -(-1)^{i-1}K_i(-2C_{\gamma_i}A_N^{(i)} + A_{N-1}^{(i)} + A_{N+1}^{(i)}) \tag{21}$$

$$E_{5N}^{(i)} = (N-1)\{\varphi_{N1}S_{N-1,\gamma_i} + \varphi_{N2}C_{N-1,\gamma_i}\}$$
$$+ (N+1)\{\varphi_{N3}S_{N+1,\gamma_i} + \varphi_{N4}C_{N+1,\gamma_i}\} = 2(-1)^{i-1}K_iS_{\gamma_i}A_N^{(i)} \tag{22}$$

where

$$K_1 = A_N^{(1)} = 0, \qquad K_2 = k/\tau^2$$

It will be observed that φ_{01} and φ_{11} appear only in the combination $\varphi_{00} = \varphi_{01} + \varphi_{11}$. This leaves an undetermined constant which can be used to assign a zero streamline.

A further analytical reduction of equations (16)-(22) is of value in the lubrication limit of small clearance. This reduction is carried out in the Appendix.

Interest in lubrication theory focuses on integrated force and moment resultants.

The resultant force per unit length on the rotating cylinder is given by (see [9])

$$X = 4\mu\pi\varphi_{04}, \qquad Y = 0 \tag{23}$$

The resultant torque referred to the cylinder center is

$$M = -4\mu\pi\tau(\varphi_{12} - \coth\gamma_1\varphi_{04}) \tag{24}$$

The shear-stress (more exactly, stress deviator) resultant is of some interest in this problem (see closing remarks in the Introduction) and is given by

$$X_s = 2\pi\mu\left\{\varphi_{04}(1 + e^{-2\gamma_1}) - \left\{\sum_{N=1}^{\infty} N[(N+1)(\varphi_{N3} + \varphi_{N+1,1})\right.\right.$$
$$+ (1-N)(\varphi_{N1} + \varphi_{N-1,3})](1 + e^{-2N\gamma_1}) + N[(N+1)(\varphi_{N4} + \varphi_{N+1,2})$$
$$\left.\left. + (1-N)(\varphi_{N2} + \varphi_{N-1,4})](1 - e^{-2N\gamma_1})\right\}\right\} \tag{25}$$

Equations (12)-(25) are formally exact. All of the physical quantities of interest and their variations may be obtained once the φ_{NM} are known. In general, the recursive relations (16)-(22) are too complicated to be solved analytically, and the φ_{NM} must be obtained numerically. This procedure has been followed in obtaining the results which are summarized in Figs. 2-8 and in the concluding section of this paper.

Important simplifications are, however, possible in the limit of our greatest interest — the lubrication (Reynolds — Sommerfeld) limit of small clearance. It is to these simplifications that we now turn.

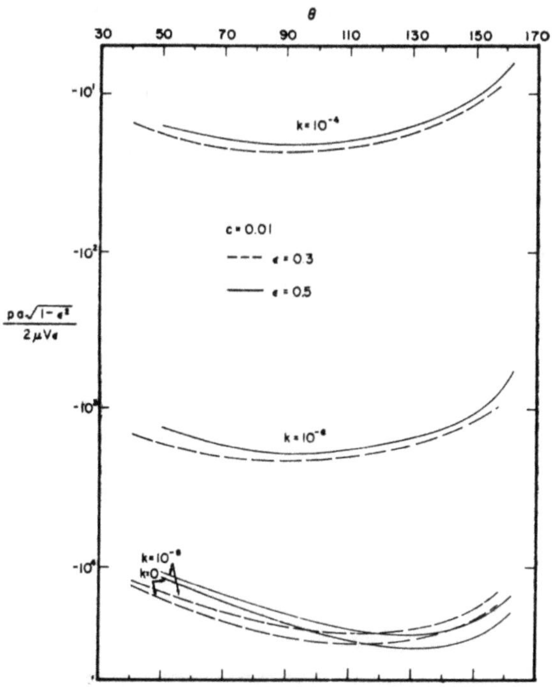

Fig. 3. Pressure variation in lubricant. Shows graphs of equation 14 [or (36)]. Pressure decreases rapidly as permeability (k) increases. Pressure minimum also shifts away from position of minimum gap ($\theta = 180°$) as permeability increases.

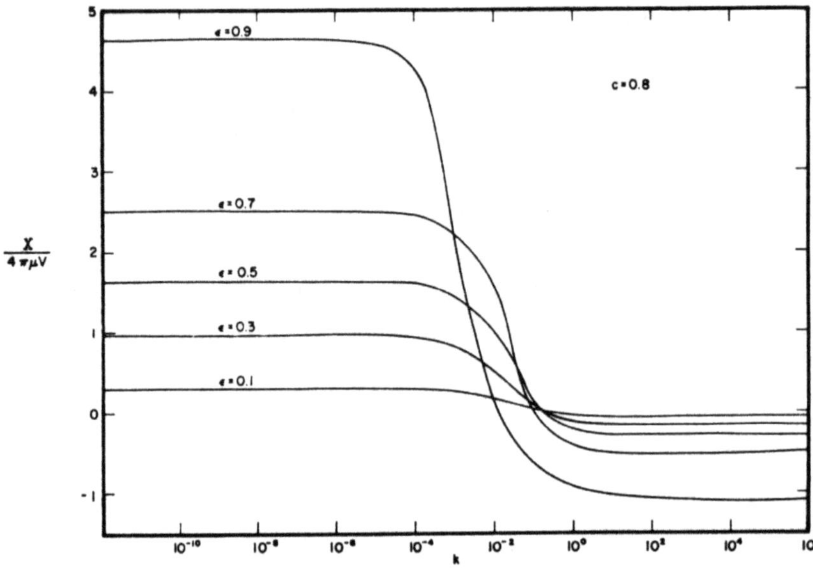

Fig. 4. Total force on journal. Shows graphs of equation (23). Total force on journal is more sensitive to changes in permeability when eccentricity is larger. For sufficiently large permeabilities, total force changes signs.

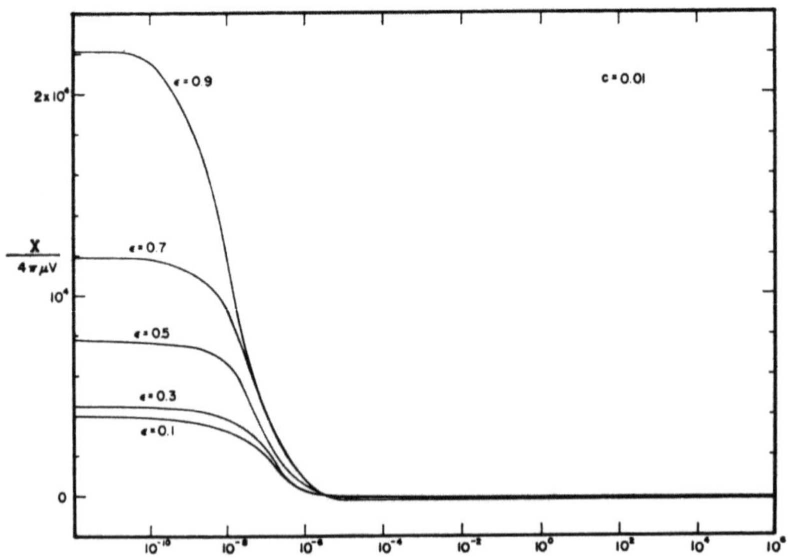

Fig. 5. Total force on journal. Shows graphs of equation (23) [or (37)] for a lubrication clearance (c = d = 0.01). Qualitatively, the situation is similar to that described under Fig. 4, except that, for smaller clearances, force is sensitive to smaller values of permeability.

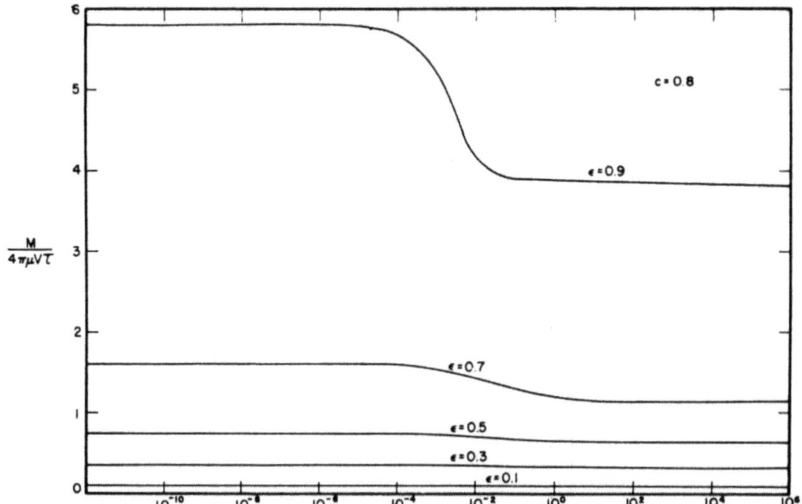

Fig. 6. Torque on journal. Shows graphs of equation (24). Torque decreased, but not significantly, for large permeability. Torque is largely determined by relative motion of bearing and journal, and this does not change as a result of bearing flow.

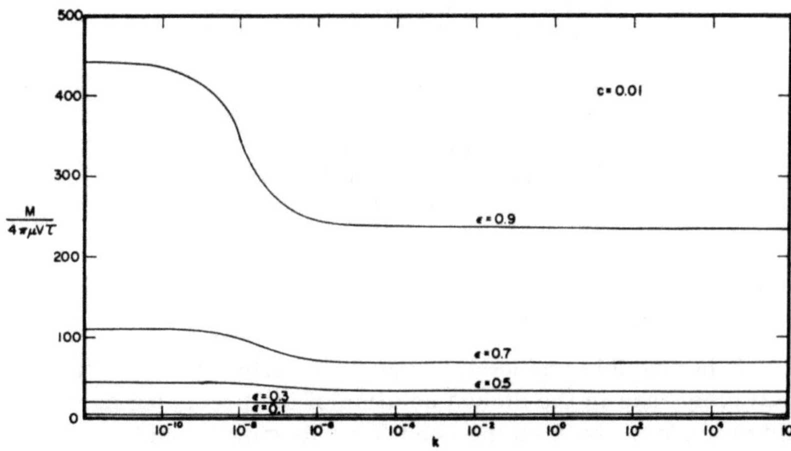

Fig. 7. Torque on journal. Shows graphs of equation (24) [or (39)] for a lubrication clearance (c = d = 0.01). Comments under Fig. 6 apply to this figure.

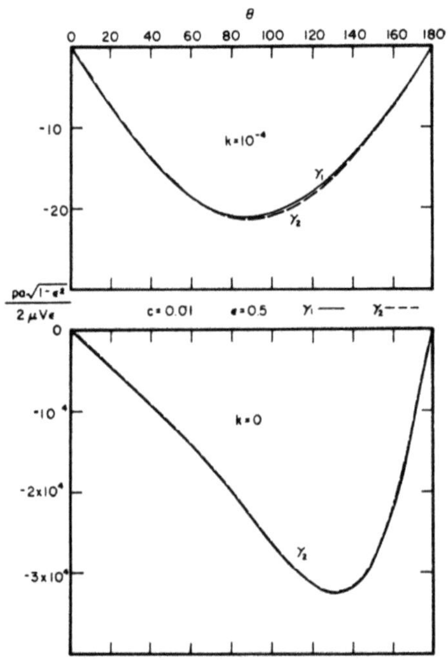

Fig. 8. Transverse pressure variation in lubricant. Shows graphs of equation (14) evaluated at journal surface (γ_1) and at bearing surface (γ_2). Difference between pressures at γ_1 and γ_2 for any fixed value of θ is a measure of sign of transverse pressure gradients (these are normally neglected in lubrication theory). It is seen that these transverse gradients are larger, but not significantly, when bearing is permeable. To neglect transverse pressure, gradients would then seem to be valid when bearing is porous.

Reynolds — Sommerfeld Limit

It is possible and useful to go to the lubrication limit of small clearance in the Stokes solution of the preceding section. This procedure obviates the need for a Reynolds equation valid in the lubrication limit. One obtains, in this way, the solution of the idealized porous-bearing lubrication problem without commitment to the appropriate Reynolds equation. Proposed Reynolds equations should then have the property that their solutions resemble the

Stokes solution when the bearing ring thickness is large. This is the procedure followed in [5] and provides one useful test for proposed approximations. We also note that the simplifications introduced in the limit of small clearance are such that it is possible to write down compact closed expressions which explicity exhibit the interplay of the small clearance, permeability, and eccentricity.

To obtain the Reynolds—Sommerfeld limit, we develop equations given in the Appendix in powers of $[\gamma]$. The lowest powers of $[\gamma]$ are retained. For higher-order coefficients ($N \geq 2$), we obtain from equations (47)-(51) the relations

$$\Delta_N = \tfrac{4}{3}N^2(N^2 - 1)[\gamma]^4 \tag{26}$$

$$\varphi_{N3} = -S_{N+1,\gamma_1}\varphi_{N4}/C_{N+1,\gamma_1} \tag{27}$$

$$\varphi_{N2} = -(N+1)C_{N-1,\gamma_1}\varphi_{N4}/(N-1)C_{N+1,\gamma_1} \tag{28}$$

$$\varphi_{N1} = (N+1)S_{N-1,\gamma_1}\varphi_{N4}/(N-1)C_{N+1,\gamma_1} \tag{29}$$

$$\Delta_N\varphi_{N4} = -4N(N-1)[\gamma]C_{N+1,\gamma_1}E_{4N}{}^{(2)} \tag{30}$$

where $E_{4N}{}^{(2)}$ is defined by equation (21) and evaluated with γ_1 replacing γ_2. Equation (30) may be expanded and equations (26)-(29) used to obtain a five-term recursion relation between $\varphi_{N-2,4}$ through $\varphi_{N+2,4}$. A method of truncation can be used to express higher-order coefficients in terms of lower-order ones. In this way, higher-order coefficients are expressed in terms of φ_{14}, φ_{13}, and φ_{04}. Equations (43), (45), and (46), which are then three equations among the three unknowns, are solved, and the result is expanded in powers of $[\gamma]$. In this last operation, care must be taken not to neglect terms of order K relative to $[\gamma]$. These operations are easily carried out for large N. Higher-order harmonics do not appreciably affect the value of the force resultants. This is especially true for k not large. For large k, the distribution of pressure is sensitive to higher harmonics, but the magnitude of the pressure approaches zero. The force coefficients are so well represented with $N = 2$ that the approximate and true solutions cannot be distinguished for moderate k and are within 10% for very large k. In the notation of lubrication theory,

$$C_{\gamma_1} = 1/\epsilon, \qquad S_{\gamma_1} = \alpha/\epsilon, \qquad \alpha = (1-\epsilon^2)^{1/2}$$

$$d = c/a, \qquad -[\gamma] = d\alpha, \qquad \tau = a\alpha/\epsilon$$

$$K = k\epsilon^2/a^2\alpha^2 = \kappa\epsilon^2/\alpha^2$$

we obtain ($N = 2$)

$$\lambda_1 = d^3\alpha^5 + 2\kappa(3 + 2\epsilon^2) \tag{31}$$

$$\lambda = \frac{d}{\alpha\epsilon^2}\left\{\frac{d^3\alpha^5}{3}(2 + \epsilon^2) + \kappa(4 - 3\epsilon^2 + 2\epsilon^4) + 6\kappa^2\epsilon^2(3 - \epsilon^2)^2/\lambda_1\right\} \tag{32}$$

$$\varphi_{04} = \frac{V}{\epsilon\alpha^2\lambda}\{d^2\alpha^5 - \kappa(4 - \epsilon^2)/\alpha + 6\kappa^2\epsilon^2(3 - \epsilon^2)/\lambda_1\} \tag{33}$$

$$\varphi_{13} = \frac{V}{\epsilon^4\lambda}\{d^2\epsilon\alpha^3 + \kappa\epsilon\alpha(2 - \epsilon^2) - 3\kappa^2(3 - \epsilon^2)/\lambda_1\} \tag{34}$$

$$\varphi_{22} = \frac{3\kappa}{\alpha\epsilon^2\lambda_1}(\epsilon^4\varphi_{13} + \epsilon^2\alpha^2\varphi_{04}/2) \tag{35}$$

$$\frac{pa}{2\mu} = \frac{-\alpha}{(\epsilon\cos\theta + 1)^2}\{\varphi_{04}\sin\theta(\epsilon\cos\theta + 1) + \varphi_{13}\epsilon^2\sin\theta + \epsilon\alpha\varphi_{22}\sin 2\theta\} \tag{36}$$

where θ is the polar angle measured from the top of the cylinder

$$X = 4\pi\mu\varphi_{04} \tag{37}$$

$$X_S = \frac{4\pi\mu V(\alpha - 1)}{\lambda\alpha^2\epsilon^3}\left\{\frac{d^3\alpha^5}{3}(1 + 2\epsilon^2 - 3\alpha) + \kappa(4\alpha^3 + \epsilon^4) - 3\kappa^2\epsilon^2(3 - \epsilon^2)(\epsilon^2 + 1)/\alpha^2\lambda_1\right\} \tag{38}$$

$$M = \frac{4\pi\mu a}{\alpha\epsilon^2\lambda}\left\{\frac{d^3\alpha^4}{3} + \kappa(8 - 15\epsilon^2 + 11\epsilon^4 - 2\epsilon^6)/2\alpha - 3\kappa\epsilon^2(3 - \epsilon^2)^2/\lambda_1\right\} \tag{39}$$

For $\kappa = 0(d^3)$ the coefficients φ_{04} and φ_{13} simplify to

$$\varphi_{13} = \varphi_{04}/\epsilon^2 = Vd^2\alpha^3/\lambda\epsilon^3 \tag{40}$$

Hence, for small clearance, the load capacity of the porous bearing is reduced by the factor $\lambda(0)/\lambda$, where $\lambda(0)$ is λ evaluated with $\kappa = 0$. The pressure distribution is then given by the compact formula

$$p = \frac{p(0)\lambda(0)}{\lambda} - \frac{6\mu Vd^2\kappa(1 - 2\epsilon^4 - \epsilon^2/2 + \epsilon^6)}{a\epsilon^6\lambda\lambda_1(\epsilon\cos\theta + 1)^2} \tag{41}$$

Here, $p(0)$ is the impermeable (Reynolds-Sommerfeld) solution. A prime effect of the bearing flow is the reduction in the amplitude of the pressure (and force resultant) by the factor $\lambda(0)/\lambda$. The distribution of pressure is also altered by the additional term in (41) but this does not affect the bearing load capacity.

Discussion of Results

Figures (2)-(8) represent the effects of the flow in the bearing on the pressure distribution and force resultants. The results are obtained from convergent machine calculations (with $a = 1$). With $d \leq 0.1$ and $0.1 \geq \epsilon \leq 0.9$, the same graphs may be obtained from equations (31)-(36).

The following observations are worth emphasis:

1. Increased flow within the porous bearing, i.e., increased permeability, leads to drastic reductions in the maximum pressure as well as a shift of this maximum away from the minimum gap, (Fig. 3). This shift is less drastic than in the porous ground [9], because the difference in the curvature of the boundaries is less.

2. The total force resultant decreases strongly with permeability (Figs. 4, 5). The decrease in the torque with permeability is much less marked (Figs. 6, 7), and the same is true for the shear-stress resultant (Fig. 2). For very large permeabilities, the total load and the (small) shear resultant are equal and negative.

3. Transverse pressure gradients vanish much less rapidly as clearance is reduced when the bearing is permeable. This can be seen by comparison of the difference in pressures at the bearing and cylinder surfaces (Fig. 8). It would appear, however, that the assumption of no transverse variation of pressure continues to be valid for the porous bearing.

Finally, it must be stressed that results obtained in this analysis are valid under the assumption that the fluid does not slip relative to the porous solid. This is a plausible but not necessarily valid assumption. The influence of the tangential component of velocity is perhaps small, in any case, but the precise nature of this boundary flow needs further investigation, particularly from an experimental point of view.

Acknowledgment

We are indebted to Mr. C. C. Shir for his work on the calculations. This work was supported, in part, by the National Science Foundation under Grants GP-3066 and GK-367 for D. D. Joseph and under GP-2258 for L. N. Tao.

APPENDIX

A further reduction of (16)-(22) is possible. Regard $E_n^{(i)}$ as given and solve (18)-(20) for the six coefficients φ_{0M}, φ_{1M}. Equations (21) and (22) are solved for the four coefficients φ_{NM}. The result is as follows:

$$[x_i] = x_2 - x_1$$

$$[x_i|y_i] = x_2y_1 - x_1y_2 \tag{42}$$

$$\Delta_0 = 2S^2_{[\gamma_i]}\{2S_{\gamma_1}S_{\gamma_2}S_{[\gamma_i]} - [\gamma_i](S_{\gamma_2}{}^2 - S_{\gamma_1}{}^2)\}$$

$$\begin{aligned}\Delta_0\varphi_{04} = {} & [E_2^{(i)}](2S_{[\gamma_i]}{}^2 - S_{[2\gamma_i]}[S_{2\gamma_i}]/2) \\ & + [S_{\gamma_i}{}^2]\{S_{[\gamma_i]}{}^2(E_1^{(1)} + E_1^{(2)}) - S_{[2\gamma_i]}[E_3^{(i)}]\}\end{aligned} \tag{43}$$

$$\Delta_0\varphi_{12} = [\gamma_i + S_{2\gamma_i}/2]\{S_{[\gamma_i]}{}^2(E_1^{(1)} + E_1^{(2)}) - S_{[2\gamma_i]}[E_3^{(i)}]\} + [S_{\gamma_i}{}^2][E_2^{(i)}]S_{[2\gamma_i]} \tag{44}$$

$$\begin{aligned}2\Delta_0\varphi_{13} = {} & [\gamma_i + S_{2\gamma_i}/2]\{[S_{2\gamma_i}/2]([E_1^{(i)}] + [C_{2\gamma_i}|E_1^{(i)}]) \\ & - 2[S_{\gamma_i}{}^2][E_3^{(i)}]\} - [S_{\gamma_i}{}^2]^2([C_{2\gamma_i}|E_1^{(i)}] - 2[E_2^{(i)}])\end{aligned} \tag{45}$$

$$\begin{aligned}2\Delta_0\varphi_{14} = {} & [\gamma_i + S_{2\gamma_i}/2]\{[S_{2\gamma_i}][E_3^{(i)}] - [S_{\gamma_i}{}^2][E_1^{(i)}] + [E_1^{(i)}|S_{2\gamma_i}][S_{2\gamma_i}/2] \\ & - [S_{\gamma_i}{}^2]\{[S_{2\gamma_i}][E_2^{(i)}] + [S_{\gamma_i}{}^2][E_1^{(i)}|S_{2\gamma_i}]\}\end{aligned} \tag{46}$$

$$\begin{aligned}\Delta_N &= 4(S_{[N\gamma_i]}{}^2 - N^2S_{[\gamma_i]}{}^2) \\ \alpha &= N - 1 \\ \beta &= N + 1\end{aligned} \tag{47}$$

$$\Delta_N\varphi_{N1} = \begin{vmatrix}\beta[E_{4N}^{(i)}C_{\beta\gamma_i}] - [E_{5N}^{(i)}S_{\beta\gamma_i}] & [S_{2N\gamma_i} - NS_{2\gamma_i}] \\ [E_{5N}^{(i)}C_{\beta\gamma_i}] - \beta[E_{4N}^{(i)}S_{\beta\gamma_i}] & [NC_{2\gamma_i} - C_{2N\gamma_i}]\end{vmatrix} \tag{48}$$

$$\Delta_N\varphi_{N2} = \begin{vmatrix}[NC_{2\gamma_i} + C_{2N\gamma_i}] & \beta[E_{4N}^{(i)}C_{\beta\gamma_i}] - [E_{5N}^{(i)}S_{\beta\gamma_i}] \\ -[NS_{2\gamma_i} + S_{2N\gamma_i}] & [E_{5N}^{(i)}C_{\beta\gamma_i}] - \beta[S_{\beta\gamma_i}E_{4N}^{(i)}]\end{vmatrix} \tag{49}$$

$$\Delta_N\varphi_{N3} = \begin{vmatrix}\alpha[C_{\alpha\gamma_i}E_{4N}^{(i)}] - [E_{5N}^{(i)}S_{\alpha\gamma_i}] & [NS_{2\gamma_i} - S_{2N\gamma_i}] \\ [C_{\alpha\gamma_i}E_{5N}^{(i)}] - \alpha[E_{4N}^{(i)}S_{\alpha\gamma_i}] & [NC_{2\gamma_i} + C_{2N\gamma_i}]\end{vmatrix} \tag{50}$$

$$\Delta_N\varphi_{N4} = \begin{vmatrix}[NC_{2\gamma_i} - C_{2N\gamma_i}] & \alpha[C_{\alpha\gamma_i}E_{4N}^{(i)}] - [E_{5N}^{(i)}S_{\alpha\gamma_i}] \\ [NS_{2\gamma_i} + S_{2N\gamma_i}] & [C_{\alpha\gamma_i}E_{5N}^{(i)}] - \alpha[E_{4N}^{(i)}S_{\alpha\gamma_i}]\end{vmatrix} \tag{51}$$

References

1. Taylor, G. I., and Miller, J C., "Fluid Flow Between Porous Rollers," Quarterly Journal of Mechanics and Applied Mathematics, London, Vol. 9, pp. 129-135, 1956.
2. Tao, L. N , and Joseph, D. D., "Fluid Flow Between Porous Rollers," Journal of Applied Mechanics, Vol. 29, Trans. ASME, Vol. 84, Series E, pp. 429-433, 1962.
3. Morgan, V. T., and Cameron, A., "Mechanism of Lubrication in Porous Metal Bearings," Proceedings, Conference on Lubrication and Wear, The Institution of Mechanical Engineers, London, 1957.
4. Cameron, A., Morgan, V., and Stainsby, A., "Critical Conditions for Hydrodynamic Lubrication of Porous Metal Bearings," Proceedings, The Institution of Mechanical Engineers, London, Vol. 176, p. 761, 1962.
5. Shir, C. C., and Joseph, D. D., "Lubrication of a Porous Bearing – Reynolds' solution," Chapter 17 of this volume.
6. Wannier, G., "A Contribution to the Hydrodynamics of lubrication," Quarterly Journal of Applied Mathematics, Vol. 8, No. 1, pp. 1-32, 1950.
7. Gheorghitza, S. I., "La formule de Stokes pour les enveloppes spheriques poreuses," Arch. Rat. Mech. Analysis, Vol. 1, No. 12, pp. 52-57, 1962.
8. Joseph, D. D., and Tao, L. N., "The Effect of Permeability on the Slow Motion of a Porous Sphere in a Viscous Liquid," ZAMM, Vol. 44, p. 361, 1964.
9. Joseph, D. D., and Tao, L. N., "Ground Flow Induced by a moving cylinder," The Physics of Fluids, Vol. 8, No. 8, pp. 1438-1449, 1965.
10. Joseph, D. D., "Note on Steady Flow Induced by Rotation of a Naturally Permeable Disk," Quarterly Journal of Mechanics and Applied Mathematics, Vol. 18, pp. 325-331, 1965.
11. Sommerfeld, A., Mechanics of Deformable Bodies, Academic Press, New York, 1950.

Chapter 17

Lubrication of a Porous Bearing—Reynolds' Solution

C. C. Shir and D. D. Joseph

Department of Aeronautics and Engineering Mechanics
University of Minnesota
Minneapolis, Minnesota

Historical Introduction and Statement of Purpose

In this report, we discuss the unique features which are introduced into the theory of hydrodynamic lubrication of bearings when the bearings are permeable and tolerate a flow of oil. The subject of the analysis is not entirely new, and a few aspects of the problem have been treated by several authors [1, 2, 4, 5, 9].

Porous bearings are extensively used in technological applications. The chief advantage of these bearings is that they require no exterior oil supply. Such bearings, after long service, are not hydrodynamically lubricated in that the load is not supported by a fluid film. Since our interest in this work is confined to hydrodynamically lubricated bearings, our analysis applies only to the initial "break-in" period (in which the oil supply is plentiful) and to the occasional pressurized system (in which oil losses are absent or annulled by a continuous supply). The bearings analyzed in this work necessarily operate at higher rotational speeds for a given load and have a potential for service in applications where light loads and high speeds are characteristic. The pressure which would develop in a solid bearing at a given journal speed is reduced when the bearing is permeable and the journal speed is unchanged. This fact has importance for applications in which lubricant cavitation is to be avoided.

The goal of this investigation is the prediction of bearing behavior as a function of bearing parameters. This is essentially a specification of the interrelations of permeability, bearing thickness, clearance, and the eccentricity of the journal. The nonpermeable counterpart of this problem is the subject of the classical theory of hydrodynamic lubrication. The bearing assembly of infinite length, with which we are concerned, is completely described for the nonpermeable bearing by the theory of Reynolds [10] and Sommerfeld [14]. There is a classical short-length bearing approximation developed by Ocvick [9] which has been extended by Morgan and Cameron [1] and Cameron, Morgan, and Stainsby [2] for applications to porous bearings. In these approximations, the circumferential variations of the pressure gradient, which may dominate the flow in longer bearings, are entirely neglected. An analysis of the porous bearing problem which starts from the low Reynolds number equation of Stokes has been constructed by Joseph and Tao [4]. This analysis, however, neglects effects of the bearing thickness. The journal is represented by a rotating circle of the bipolar family, and the inner bearing surface, by another circle of this family. The bearing itself is presumed infinite in extent and models, as is shown in the sequel, a thick bearing (radius ratio $>3/2$).

Our present effort starts from a Reynolds equation modified to account for the through-flow (as suggested by Cameron et al. [1, 2]) and matched to the bearing flow by matching conditions at the common boundary. The bearing itself is presumed to be encased in a solid so that the outer bearing surface is impermeable. The bearing has a finite thickness and is infinite in length. The pressure in the bearing is harmonic, and its first derivative (which gives the normal velocity) vanishes at the outer solid-bearing surface. At the inner bearing surface, the harmonic functions must satisfy the modified Reynolds equation.

The results of this investigation support conclusions of [4] and extend these to accommodate the effects of the finite thickness of the porous bearing. The chief effect of the permeability is the reduction of pressures which would otherwise develop. The force resultants are lower as a result of the diminution of the pressure, and the shear-stress resultant may be nonnegligible relative to the pressure-force resultant. Effects of changing the thickness of the bearing are pronounced as the bearing thickness is small, and they rapidly diminish as the ratio of the outer to inner bearing is increased beyond 3/2.

Statement of Problem, Including Mathematical Formulation

Consider a solid journal rotating in a porous bearing of a finite constant thickness contained in a solid housing with end effects neglected. The cross sections of the journal and the bearing are nonconcentric circles (Fig. 1).

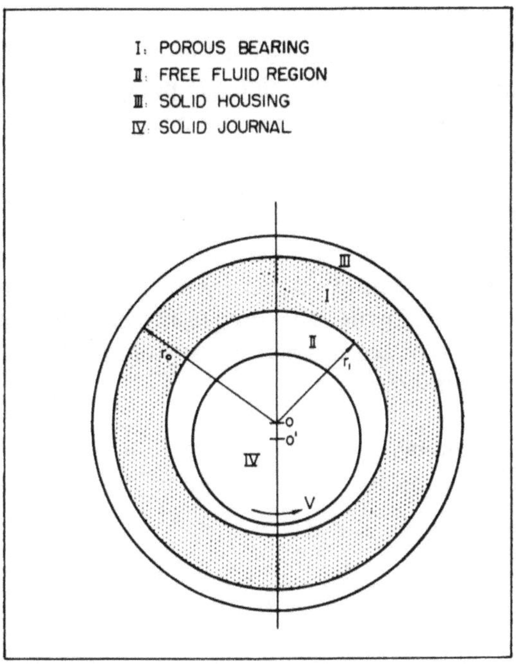

Fig. 1. Bearing configuration.

The assumptions of conventional lubrication theory retained in this analysis are:

1. The lubricant is Newtonian and incompressible.
2. The film is so thin, compared with the ratio of kinematic viscosity to linear velocity, that the motion of the fluids is laminar, inertial terms and bearing curvature can be neglected, and pressure and viscosity are uniform across the film.
3. No slip occurs between fluid and bearing surfaces.
4. There is no end leakage from the bearing (problem is two-dimensional).

The motion of the viscous fluid is induced by the rotation of the journal. The behavior of the fluid outside the porous bearing is obviously influenced by the flow within the porous bearing. The solutions to the governing equations valid in different adjacent regions must be matched on the boundaries.

Briefly, the problem is characterized by the requirements that the flow satisfy:

1. The appropriate specialization of the Reynolds equation in the free region (outside the porous region).
2. The Laplace equation for the pressure in the porous region.
3. No-slip conditions immediately outside the porous region (see [4] for full discussion).
4. A condition requiring the radial filter velocity to vanish on the outer (solid) boundary of the porous bearing.
5. Continuity of pressure across the porous bearing.
6. Continuity of the normal component of the velocity across the porous boundary.

In the porous-bearing ring (assumed homogenous and isotropic), the filter velocity $\mathbf{Q}$ is related to the relative pressure $\hat{p}$ by Darcy's law

$$\mathbf{Q} = \frac{k}{\mu} \nabla \hat{p} \tag{1}$$

From mass conservation, $\mathbf{Q}$ is solenoidal, implying that

$$\nabla^2 \hat{p} = 0 \tag{2}$$

It is required that $\hat{p}$ be a single-valued harmonic function, the normal derivative of which vanishes on the outer radius of the ring and which satisfies a Reynolds equation along the inner circumference of the ring.

We next follow Morgan and Cameron [1] in a derivation of Reynolds' equation modified to account for mass exchanges with the oil-saturated bearing. As is usual in lubrication theory, we unwrap the gap between journal and bearing. It is assumed then that lengthwise variations are dominated by transverse variations so that the equation

$$u = \frac{1}{2\mu}\frac{dp}{dx}(y^2 - yh) + \frac{V}{h}y \tag{3}$$

where $h(x)$ is the local channel height, is locally valid. Conservation of mass requires that (Fig. 2).

$$Q_2 - Q_1 + Q_0 = 0 \tag{4}$$

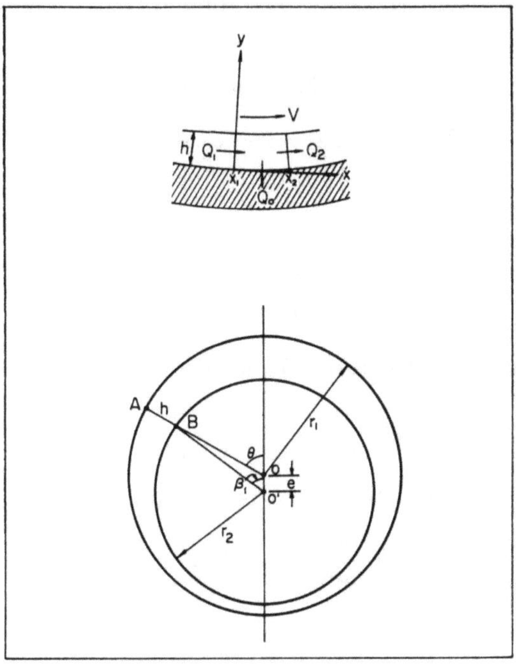

Fig. 2. Geometry of fluid film.

where Q_0 is given by the Darcy law

$$Q_0 = -\left(\frac{k}{\mu}\frac{\partial \hat{p}}{\partial r}\right)_{r_1} \Delta x \tag{5}$$

From equations (4) and (5), it follows that

$$\frac{dQ}{dx} - \left(\frac{k}{\mu}\frac{\partial \hat{p}}{\partial r}\right)_{r_1} = 0 \tag{6}$$

where

$$Q = \int_0^{h(x)} u dy = -\frac{1}{12\mu}\frac{dp}{dx}h^3 + \frac{V}{2}h \tag{7}$$

Now let $x = r_1\theta$ and combine (6) and (7) to obtain

$$\frac{d}{d\theta}\left(h^3\frac{dp}{d\theta}\right) - 6\mu V r_1 \frac{dh}{d\theta} + \left(12kr_1^2\frac{\partial \hat{p}}{\partial r}\right)_{r_1} = 0 \tag{8}$$

This is a Reynolds equation with an additional term to accommodate the mass exchange between the film and bearing. The usual approximation $h(\theta) = c(1 + \epsilon \cos\theta)$ for small clearances is valid in the present context. The term $dh/d\theta = -\epsilon c \sin\theta$ is a driving term for (8).

It is convenient to introduce the following dimensionless variables:

$$\begin{aligned} \eta &= h/e = 1 + \epsilon\cos\theta \\ K &= 12kr_1/c^3 \\ \beta &= r_0/r_1 \\ R &= r/r_1 \\ P &= pc^2/6\mu r_1 V \\ \hat{P} &= \hat{p}c^2/6\mu r_1 V \end{aligned} \tag{9}$$

Using the variables defined in (9), equation (8) becomes

$$\frac{d}{d\theta}\left(\eta^3\frac{dP}{d\theta}\right) - \frac{d\eta}{d\theta} + \left(K\frac{\partial \hat{P}}{\partial R}\right)_1 \tag{10}$$

The quantity $\phi = kr_1/c^3$ ($K = 12\phi$) is a design factor. It follows that

$$P = \frac{pc^2}{6\mu r_1 V} = F(\theta;\ \phi, \epsilon, \beta)$$

and

$$\hat{P} = \frac{\hat{p}c^2}{6\mu r_1 V} = \hat{F}(r, \theta;\ \phi, \epsilon, \beta)$$

The parameters (ϕ, ϵ, β) characterize the solution.

Solutions

In this section, we develop a formally exact solution and the approximations which make the solution computationally useful. We first note that equation (8) and the ring boundary conditions imply that p and $\hat{p}$ are antisymmetric. Harmonic and antisymmetric $\hat{p}$ with a vanishing normal derivative at $r = r_0$ is given by

$$\hat{p}\,(r, \theta) = \sum_{n=1}^{\infty} B_n \left(\frac{r_1}{r}\right)^n \left[1 + \left(\frac{r}{r_0}\right)^{2n}\right] \sin n\theta \tag{11}$$

The continuity of pressure implies $p(r_1, \theta) = \hat{p}(\theta)$. This reduces our problem to finding B_n compatible with equation (8). We first substitute harmonic

$$\hat{p}(r_1, \theta) = \sum_{n=1} B_n C_n \sin(n\theta) \tag{12}$$

where

$$C_n = 1 + \beta^{-2n},\ \beta = r_0/r_1,\ D = \frac{6\mu r_1 V}{c^2}$$

into

$$\frac{d}{d\theta}\left(\eta^3 \frac{d\hat{p}}{d\theta}(r_1, \theta)\right) - D\frac{d\eta}{d\theta} + \left(K r_1 \frac{\partial \hat{p}}{\partial r}\right)_{r_1} = 0 \tag{13}$$

to find a seven-term recursion formula for the coefficients B_n:

$$[(A_0 + A_2)C_1 + Kd_1]B_1 + 2\,(A_1 + A_3)C_2B_2 + 3A_2C_3B_3 + 4A_3C_4B_4 = D\epsilon \tag{14}$$

$$(A_1 + A_3)C_1B_1 + (2C_2A_0 + Kd_2)B_2 + 3C_3A_1B_3 + 4C_4A_2B_4 + 5C_5A_3B_5 = 0 \tag{15}$$

$$A_2C_1B_1 + 2A_1C_2B_2 + (3C_3A_0 + Kd_3)B_3 + 4A_1C_4B_4 + 5A_2C_5B_5 + 6A_3C_6B_6 = 0 \tag{16}$$

$$\begin{aligned}(m-3)A_3C_{m-3}B_{m-3} &+ (m-2)A_2C_{m-2}B_{m-2} + (m-1)A_1C_{m-1}B_{m-1} \\ &+ (mC_mA_0 + Kd_m)B_m + (m+1)A_1C_{m+1}B_{m+1} + (m+2)A_2C_{m+2}B_{m+2} \\ &+ (m+3)A_3d_{m+3}B_{m+3} = 0 \qquad (m = 4, 5, 6, \ldots)\end{aligned} \tag{17}$$

where

$$\begin{aligned} &C_m = 1 + \beta^{-2m},\ d_m = 1 - \beta^{-2m},\ (m = 1, 2, \ldots) \\ &A_0 = 1 + \tfrac{3}{2}\epsilon^2,\ A_1 = \tfrac{3}{2}\epsilon(1 + \tfrac{1}{4}\epsilon^2) \\ &A_2 = \tfrac{3}{4}\epsilon^2,\ A_3 = \tfrac{1}{8}\epsilon^3 \end{aligned} \tag{18}$$

To obtain the coefficients B_n, it is necessary to solve an infinite number of the linear equations. The infinite system is ap-

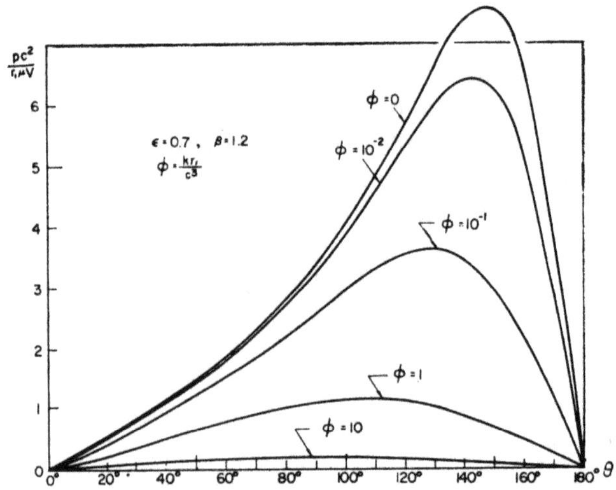

Fig. 3. Pressure distribution in lubricant for fixed eccentricity (ϵ) and radius ratio (β). Pressure is a decreasing function of permeability parameter (ϕ). Pressure maximum decreases and shifts away from position of minimum gap ($\theta = 180°$) as permeability parameter increases.

proximated with a truncated system. The coefficients B_n ($n > 10$) are set equal to zero, and the 10 linear equations are solved for the 10 unknown B_n ($n \leq 10$). It develops, as will be seen in the sequel, that the true solution is very adequately represented with n less than 10. For $\epsilon \leq 0.5$, three terms reasonably represent the true solution. For ϕ very large, only one term is enough. The result for $n = 3$, $B_n = 0$ ($n > 3$) is as follows:

$$\Delta = K^3 d_1 d_2 d_3 + K^2 \left[C_1 d_2 d_3 \left(1 + \frac{9}{4}\epsilon^2\right) + 2C_2 d_1 d_3 \left(1 + \frac{3}{2}\epsilon^2\right) \right.$$

$$\left. + 3C_3 d_1 d_2 \left(1 + \frac{3}{2}\epsilon^2\right)\right] + K \left[6d_1 C_2 C_3 \left(1 + \frac{3}{4}\epsilon^2 + \frac{9}{8}\epsilon^4\right) \right.$$

$$\left. + 3d_2 C_1 C_3 \left(1 + \frac{5}{4}\epsilon^2 + \frac{45}{16}\epsilon^4\right) + 2d_3 C_1 C_2 \left(1 + 6\epsilon^2 + \frac{39}{8}\epsilon^4\right)\right]$$

$$+ 6C_1 C_2 C_3 \left(1 + \frac{3}{4}\epsilon^2 + \frac{3}{4}\epsilon^4\right) \tag{19}$$

$$B_1 = \frac{D\epsilon}{\Delta}\left[K^2 d_2 d_3 + K\left(1 + \frac{3}{2}\epsilon^2\right)(3d_2C_3 + 2C_2d_3) + 6C_2C_3\left(1 + \frac{3}{4}\epsilon^2 + \frac{9}{8}\epsilon^4\right)\right] \tag{20}$$

$$B_2 = -\frac{D\epsilon^2}{2\Delta}\left[KC_1d_3(3 + \epsilon^2) + 3C_1C_3\left(3 + \frac{13}{4}\epsilon^2 - \frac{15}{16}\epsilon^4\right)\right] \tag{21}$$

$$B_3 = \frac{3D\epsilon^3}{4\Delta}\left[KC_1d_2 + \frac{1}{2}C_1C_2(8 + \epsilon^2 + \epsilon^4)\right] \tag{22}$$

This solution and the true solution are indistinguishable for $\epsilon \leq 0.5$ (Fig. 12).

It follows that

$$B_n = 6\mu r_1 V/c^2 f_n\ (kr_1/c^3,\ {}_0r/r_1,\ \epsilon) \tag{23}$$

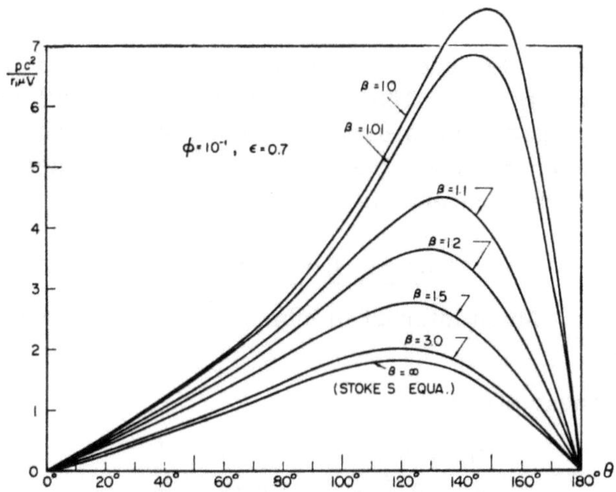

Fig. 4. Pressure distribution in lubricant for fixed values of eccentricity (ϵ) and permeability parameter (ϕ). Pressure maximum decreases and shifts away from position of minimum gap as radius ratio increases. Rate of change of pressure is greatest for a radius ratio of 1 and decreases as radius ratio increases. For very large values of radius ratio ($\beta \to \infty$), this solution tends to the solution of Stokes' equation (4) for the same flow.

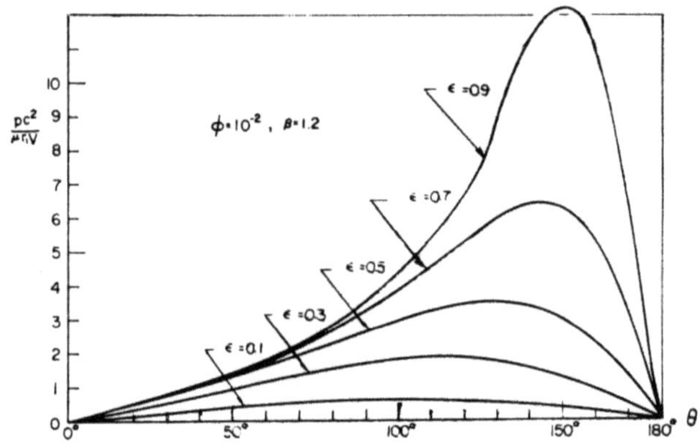

Fig. 5. Pressure distribution in lubricant for fixed values of radius ratio (β) and permeability parameter (ϕ). Pressure magnitude is a decreasing function of eccentricity as in the impermeable case.

Since

$$P = \sum_{n=1}^{\infty} f_n\,(kr_1/c^3,\ r_0/r_1,\ \epsilon)\sin n\theta \tag{24}$$

our solution is independent of c except as changes in c induce changes in $\phi = kr_1/c^3$. The true pressure scales with $\mu r_1 V/c^2$, except as changes in pressure are induced by permeability (through ϕ), and these changes which depend on c^{-3} become increasingly important as the clearance is reduced.

Force Resultants

Torque. Much interest in lubrication theory focuses on the resultants.

The shear stress is

$$\frac{\tau_0}{\mu} = \left(\frac{\partial u}{\partial y}\right)_{y=0} = -\frac{h}{2\mu}\frac{dp}{dx} + \frac{V}{h} \tag{25}$$

$$\frac{\tau_1}{\mu} = \left(\frac{\partial u}{\partial y}\right)_{y=h} = \frac{h}{2\mu}\frac{dp}{dx} + \frac{V}{h} \tag{26}$$

where

$$\frac{dp}{dx} = \frac{1}{r_1}\frac{dp}{d\theta} = \frac{1}{r_1}\sum_{n=1}^{\infty} B_n C_n n\cos n\theta$$

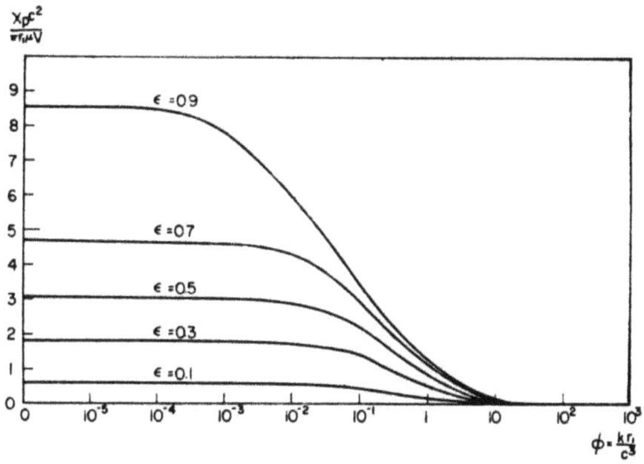

Fig. 6. Resultant pressure force on journal as a function of permeability (ϕ) with eccentricity (ϵ) as a parameter for a fixed radius ratio (β = 1.2). Pressure is more pronounced for larger values of eccentricity.

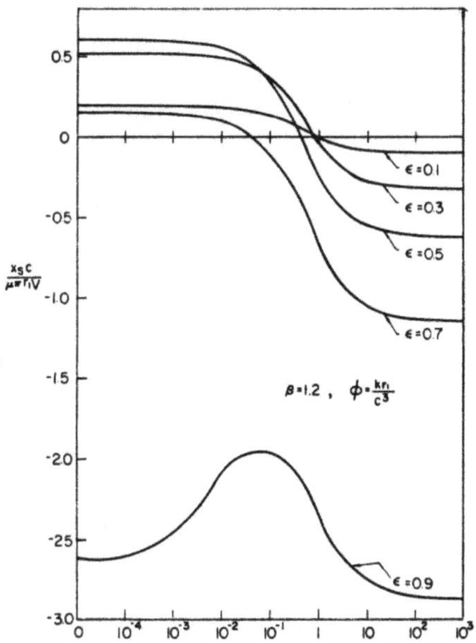

Fig. 7. Resultant shear force on Journal as a function of permeability (ϕ) with eccentricity (ϵ) as a parameter for a fixed radius ratio (β = 1.2). Shear resultant is always negative for sufficiently large values of ϕ or ϵ. For large ϕ, shear resultant may be of some order of magnitude as pressure resultant.

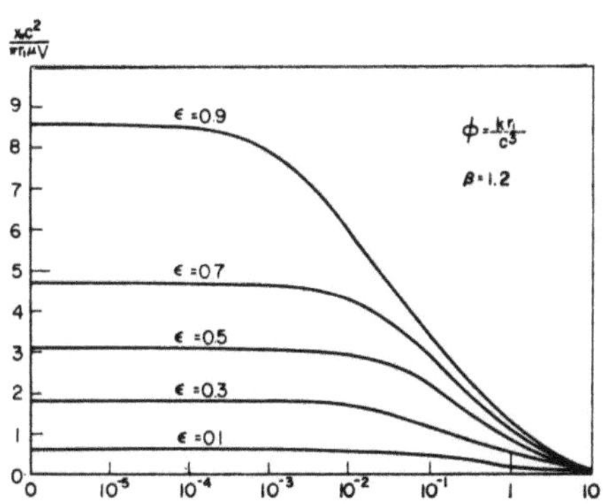

Fig. 8. Resultant total force on journal as a function of permeability (ϕ) with eccentricity (ϵ) as a parameter for a fixed radius ratio (β = 1.2). Total force is a combination of pressure resultant (Fig. 6), which scales with C^{-2} for small C, and shear stress resultant (Fig. 7), which scales C^{-1} for small C. For small permeabilities, pressure resultant dominates.

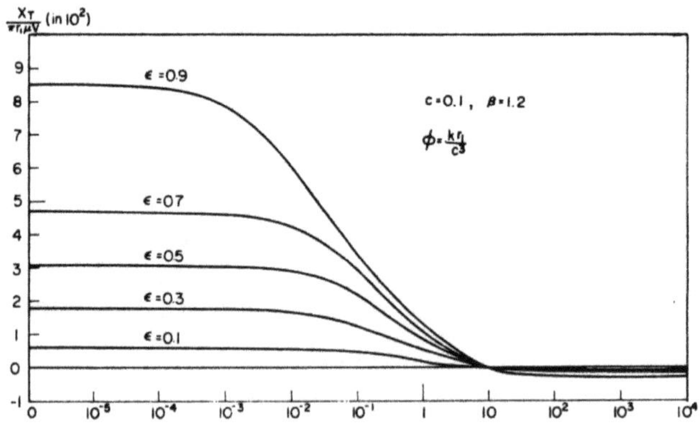

Fig. 9. Resultant total force on journal as a function of permeability (ϕ) with eccentricity (ε) as a parameter for fixed values (β = 1.2, C = 0.1). For large ϕ, force is dominated by shear resultant, which is negative. With r_1 = 1 in. and C = 0.001 in., ϕ = 10 implies k = 10^{-9} sq. in., which is a very high permeability and not at all typical of porous bearing.

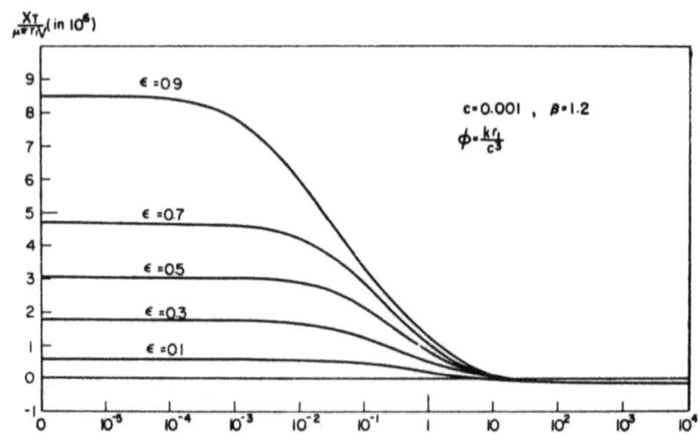

Fig. 10. Resultant total force on journal. This is qualitatively the same as Fig. 9. With r_1 = 1 in. and C = 0.001 in., ϕ = 10 implies k = 10^{-9} in.^{2}, which is typical for porous bearing.

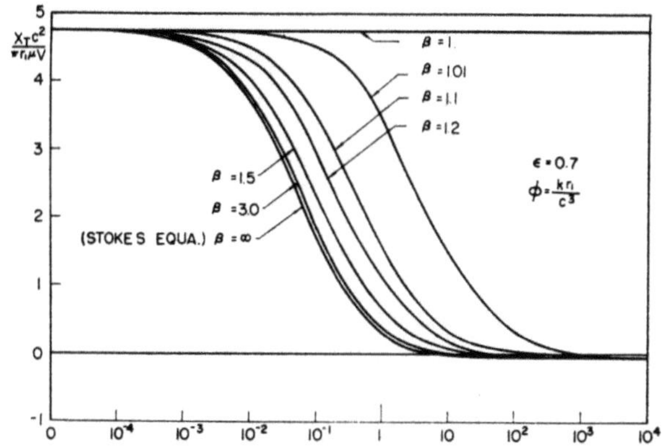

Fig. 11. Resultant total force on journal as a function of permeability (ϕ) with radius ratio (β) as a parameter and a fixed value of eccentricity (ϵ = 0.7). Total force decreases with decreasing rapidly as radius ratio is increased from 1. Value of β = 1 corresponds to an impermeable bearing, and $\beta \rightarrow \infty$ to an infinite porous matrix. In this latter limit, solution may be compared with that given in [4] for same geometry. Solution in [4] starts from two-dimensional equations of Stokes. Reynolds' and Stokes' solutions are nearly identical.

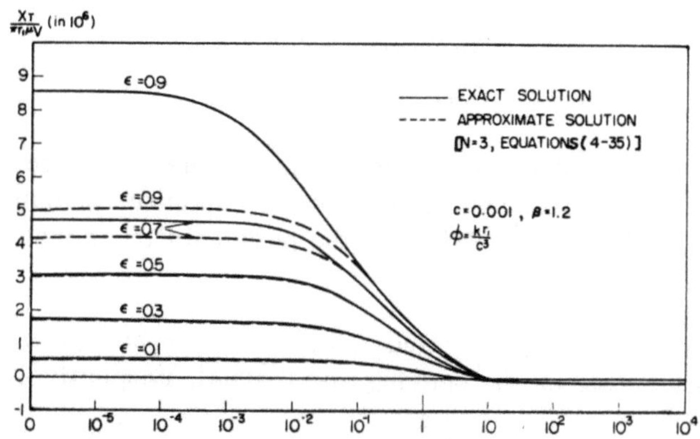

Fig. 12. Resultant total force on journal. The approximate solution, equations (19) to (22), is compared with numorical solution of equations (15) to (18). Approximate solutions are less accurate for larger valuss of eccentricity.

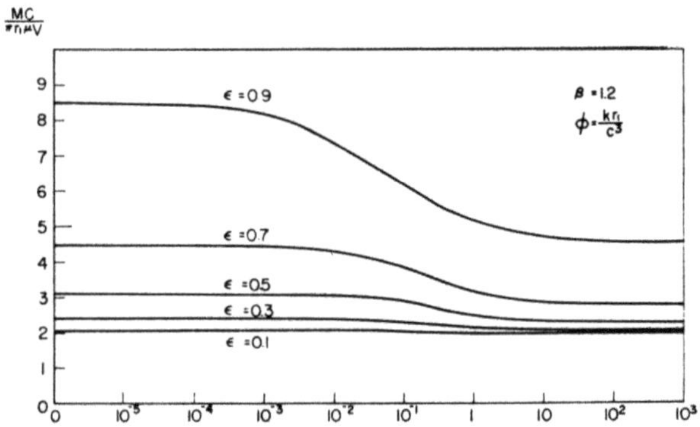

Fig. 13. Resultant moment on journal as a function of permeability (ϕ) with eccentricity as a parameter and for fixed value of radius ration(β = 1.2). Moment decreases, but not significantly, for large permeability. Torque is largely determined by relative motion of bearing and journal, and this does not change as a result of bearing flow.

The torques acting on the journal and bearing are M_1 and M_0:

$$M_i = r_1^2 \int_0^{2\pi} \tau_i d\theta; \qquad i = 0,1 \tag{27}$$

$$\frac{M_1}{\mu\pi r_1^2} = \frac{2V}{c(1-\epsilon^2)^{1/2}} + \frac{c\epsilon C_1 B_1}{2\mu r_1} \tag{28}$$

$$\frac{M_0}{\mu\pi r_1^2} = \frac{2V}{c(1-\epsilon^2)^{1/2}} - \frac{c\epsilon C_1 B_1}{2\mu r_1} \tag{29}$$

$$M_1 - M_0 = \pi c\epsilon C_1 B_1 r_1 \tag{30}$$

It is seen that the permeability and the eccentricity are the sources of the difference of the torques upon the bearing and journal. The difference vanishes as the permeability increases and the eccentricity diminishes.

Force. $X_{0,T}$, $X_{1,T}$ and $Y_{0,T}$, $Y_{1,T}$ are the x, y components of forces acting on the bearing and journal, respectively:

$$X_{1,T} = \int_0^{2\pi} (p \sin\theta + \tau_1 \cos\theta)\, r_1 d\theta \tag{31}$$

$$X_{0,T} = \int_0^{2\pi} (p \sin\theta + \tau_0 \cos\theta) r_1 d\theta \tag{32}$$

$$\frac{X_{1,T}}{\pi r_1} = B_1C_1\left(1 + \frac{c}{2r_1}\right) + B_2C_2\frac{c\epsilon}{2r_1} + \frac{\mu V}{c}\,\frac{2}{\epsilon}\left(1 - \frac{1}{(1-\epsilon^2)^{1/2}}\right) \tag{33}$$

$$\frac{X_{0,T}}{\pi r_1} = B_1C_1\left(1 - \frac{c}{2r_1}\right) - B_2C_2\frac{c\epsilon}{2r_1} + \frac{\mu V}{c}\,\frac{2}{\epsilon}\left(1 - \frac{1}{(1-\epsilon^2)^{1/2}}\right) \tag{34}$$

$$Y_{1,T} = \int_0^{2\pi} (p\cos\theta + \tau_1\sin\theta)r_1 d\theta \tag{35}$$

$$Y_{0,T} = \int_0^{2\pi} (p\cos\theta + \tau_0\sin\theta)r_1 d\theta \tag{36}$$

The pressure force and shear-stress resultant, respectively, are

$$\frac{X_{1,p}}{\pi r_1} = \frac{X_{0,p}}{\pi r_1} = B_1C_1 \tag{37}$$

$$\frac{X_{1,s}}{\pi r_1} = B_1C_1\frac{c}{2r_1} + B_2C_2\frac{c\epsilon}{2r_1} + \frac{\mu V}{c}\,\frac{2}{\epsilon}\left(1 - \frac{1}{(1-\epsilon^2)^{1/2}}\right) \tag{38}$$

$$\frac{X_{0,s}}{\pi r_1} = -B_1C_1\frac{c}{2r_1} - B_2C_2\frac{c\epsilon}{2r_1} + \frac{\mu V}{c}\,\frac{2}{\epsilon}\left(1 - \frac{1}{(1-\epsilon^2)^{1/2}}\right) \tag{39}$$

For small permeability, the pressure force dominates, and the shear-stress resultant is neglected as small compared with the pressure resultant. But the pressure resultant decreases markedly with permeability (Fig. 6) and the shear stress resultant decreases only slightly and turns negative (Fig. 7). The total force resultant decreases strongly and turns negative as the shear stress resultant becomes dominant (Figs. 9, 10).

For the impermeable case, the shear stress resultant, which is pointed out in [4], changes sign at $\epsilon = 0.721$ and is negative for large values of ϵ. Hence, for large eccentric ratios, the bearing capacity can vanish and even change direction.

From equation (28), (29), (37), (38), (39) and the fact that

$$\frac{B_n c^2}{6\mu r_1 V} = f_n(\phi, \beta, \epsilon) \tag{40}$$

one may infer that

$$\frac{X_p c^2}{\mu r_1 V},\ \frac{X_s c}{\mu r_1 V},\ \text{and}\ \frac{Mc}{\mu r_1 V}$$

depend only on the parameters (ϕ, β, ϵ). Since X_P dominates X_T when ϕ is small, and X_S dominates X_T when ϕ is large, it follows that

$$\frac{X_T c^2}{\mu r_1 V} \text{ or } \frac{X_T c}{\mu r_1 V}$$

depend only on (ϕ, β, ϵ) as ϕ is small or large (Fig. 8). In fact, the numerical results indicate that small is approximately $\phi \leq 10$ and large is $\phi > 10$.

Discussion on Solutions

The Fourier series which is formally exact will represent the true solution if coefficients B_n can be found which reduce the linear equation to identities. This system is infinite, and its exact convergence characteristics are not easily examined. Suffice it to say that the truncated system does rapidly converge. With $\epsilon \leq 0.9$, the solutions obtained by truncating the system with $n > 10$ do not differ from one another within the error (10^{-4}) represented by our graphical representation of the results.

The Fourier series would appear to converge for all values of ϕ and rapidly as ϕ is large. For large values of ϕ (≥ 10), only one term is enough. As $\phi \to \infty$ $(k \to \infty$ or $c \to 0)$, the coefficient $B_1 \sim 0(1/\phi)$, $B_j \sim 0(1/\phi^2)$, $j = 2, 3, \ldots$. The pressure dies like $1/\phi$ in this limit, but the velocities in lubricant and bearing are finite. In general, the Fourier series converges more rapidly as ϵ is small, with a uniform pressure resulting in the limit $\epsilon \to 0$, as is consistent with the physics of flow.

As $\epsilon \to 1$, the series converges slowly, while, if $\phi = 0$, the solution diverges. In this case, the journal and the bearing are in contact, and fluid cannot be carried through the gap.

It is of some interest that Reynolds himself solved the corresponding impermeable problem with a Fourier series [10]. However, the Fourier coefficients A_n were expanded in a power series in ϵ,

$$A_n = \sum_{k-1} C_{k,n}\, \epsilon^k$$

and the interval of convergence of the power series is $\epsilon < 0.6$. It is not the Fourier series which diverges, as is sometimes stated in the lubrication literature [13, p. 178], but the power series in ϵ.

Conclusions

Figures 3-13 represent the effects of the flow in the porous bearing on the pressure distribution and force resultants. The results are obtained from machine calculations with $r_1 = 1$ and $0.1 \leq \epsilon \leq 0.9$. The important parameters are the clearance, eccentricity, permeability, and radius ratio.

The following observations are worth emphasis:

1. The effect of increasing the permeability, other parameters fixed, is to reduce the pressure, shift the maximum (Fig. 3), and to reduce the magnitude of the total-force resultant (Figs. 8, 9, 10) and the pressure-force resultant (Fig. 6). The effects on the torque and shear stress of increases in permeability are less significant (Figs. 7 and 13).
2. The effects of permeability are more pronounced when:

(a) The clearance is small. The dependence of the pressure on permeability is of the form

$$p \propto (\mu r_1 V/c^2)\, F\, (\phi, \beta, \epsilon)$$

Thus the diminution of c enhances the effect of the permeability parameter even more strongly than the corresponding direct effect on the pressure. The permeability enters the problem through ϕ. The permeability parameter can always be made significant by reducing the clearance (Fig. 3).

(b) The eccentric ratio is near unity. An eccentric ratio of 1 implies contact, and our solutions are not valid in this limit. The essential effects are easily read from Figs. 8, 9, 10, and 13. These graphs show that the variation of total force and torque with permeability is much more pronounced as the values of ϵ approaches unity. The range of clearance and eccentricity in which small permeability is important may be read from the graphs.

3. The radius ratio can, of course, have an effect only if the bearing is permeable. In the limit $\beta = r_0/r_1 = 1$, there is no porous matrix, and the impermeable solution applies. In the limit $\beta \to \infty$, the matrix is an infinite porous space with an internal hole in which the journal rotates. There is a solution available for this latter case which was derived from the slow-motion equations of Stokes [4]. Our solution and that of [4] are compared in Figs. 4 and 11. The agreement in all cases is excellent.

The variation of the pressure distribution and force resultant with β is shown in Figs. 4 and 11, respectively. The zero-thickness solution is consistent with Sommerfeld's solution, as was to be expected. It is of importance that the effect of increase in the radius is much more pronounced as the thickness approaches zero. The difference between the pressure distribution in an infinite porous space and for a radius ratio of 1.5 is less than 20%. This tends to establish the solutions given in [4] as appropriate to thick bearings ($\beta > 1.5$).

4. The force resultants decrease with increasing permeability. The pressure resultant tends to zero (Fig. 6). This is not true of the shear-stress resultant (Fig. 7), which turns negative for sufficiently large ϵ. The total force, which is the sum of the pressure and shear-stress resultant, turns negative with the shear resultant when the permeability effects have reduced the magnitude of the pressure (Figs. 9 and 10 and cf. [4]).

5. The journal may run at a higher speed for a fixed pressure when the bearing is permeable. If this pressure is the cavitation limit, then a higher speed may be obtained by using a porous bearing. From Fig. 3 ($\epsilon = 0.7$, $\beta = 1.2$), the pressure parameter $pc^2/\mu r_1 V$ is reduced to half of that in the nonpermeable case when $\phi = 10^{-1}$, to 1/5 when $\phi = 1$, to 1/30 when $\phi = 10$. Thus, for a fixed $\hat{p}$, we may increase the journal speed by a factor of 2 when $\phi = 10^{-1}$, a factor of 5 when $\phi = 1$, and a factor of 30 when $\phi = 10$.

Acknowledgment

This work was supported in part by the National Science Foundation under Grants GP-3066 and GK-367.

References

1. Cameron, A. and Morgan, V., "Mechanism of Lubrication in Porous Metal Bearings," Proceedings, Conference on Lubrication and Wear, The Institution of Mechanical Engineers, London, 1957.
2. Cameron, A., Morgan, V., and Stainsby, A., Proceedings, The Institution of Mechanical Engineers, London, England, Vol. 176, p. 761, 1962.
3. Hersey, M., Theory of Lubrication, John Wiley & Sons, New York, 1936.
4. Joseph, D. D., and Tao, L. N., "Lubrication of a porous bearing — Stokes' solution," Chapter 16 of this volume.

5. Tao, L. N., and Joseph, D. D., "Fluid Flow Between Porous Rollers," Journal of Applied Mechanics, Vol. 29, Trans. ASME, Vol. 84, Series E, pp. 429-433, 1962.
6. Joseph, D. D., and Tao, L. N., "The Effects of Permeability on the Slow Motion of a Porous Sphere in a Viscous Liquid," ZAMM, Vol. 44, No. 8/9, 1964.
7. Muskat, M., Flow of Inhomogeneous Fluid, McGraw-Hill, New York, 1937.
8. Norton, A., Lubrication, McGraw-Hill, New York, 1942.
9. Ocvick, F., and Dubois, G., NACA Technical Notes Nos. 2808 and 2809, 1952.
10. Reynolds, "On the Theory of Lubrication and Application," Philosophical Transactions of the Royal Society of London, London, Vol. 177, 1886.
11. Scheidegger, A., The Physics of Flow Through Porous Media, Macmillan, New York, 1957.
12. Slaymaker, R., Bearing Lubrication Analysis, John Wiley & Sons, New York 1955.
13. Shaw, C., and Macks, E., Analysis and Lubrication of Bearing, McGraw-Hill, New York, 1949.
14. Sommerfeld, A., Mechanics of Deformable Bodies, Academic Press, New York, 1950.

Chapter 18

Selecting the Right Lubricant for Self-Lubricating Bearings and Parts

James J. Scott

Bound Brook Bearing Corporation of America
Bound Brook, New Jersey

Self-lubricating bearings and parts fill an old and constantly growing field, but very little has been published on the most important constituent of these materials – the lubricants with which they are impregnated. In fact, aside from stopping and starting, it is relatively unimportant what bearing material is used, or how well it is sintered, as long as the lubricant is functional. Since most units are subject to many stops and starts, it is difficult to surpass bronze bearings, and so all the data we will discuss are based on ASTM B202 Grade 1 Class A material, but the basic information could apply to other materials and self-lubricating parts.

Since there is very little lubricant in these bearings, it is very important that this lubricant be stable and have good lubricity. As an example of this small amount of lubricant, the popular $\frac{1}{2}$ in. I.D. × $\frac{5}{8}$ in. O.D. × 1 in. long bearing has only twelve drops of oil and this bearing is used in automobiles, appliances, and small tools, generally without any added lubricant. The oil is expected

Table 1. Recommended Oils

Property	Light	Medium	Heavy
Viscosity–SSU @ 100°F	150	300	750
Viscosity–SSU @ 210°F	44	54	79
Viscosity index	112	110	102
Pour point	10°F	12°F	15°F
D 943 Oxidation test	3000 hr	3000 hr	2000 hr
Approx. SAE Grade	10	20	40

to stay fluid, lubricate, and not break down into sludge or varnish.

Recommended Viscosities

In Table 1, we list three grades of lubricants that will satisfy practically all applications, as has been proven over many years in thousands of installations. We insist on an oil with a high viscosity index, which is an indication of its ability to resist thinning out with increasing temperature. 100 used to be considered the apex, but with special additives, not always stable in themselves, 100 is no longer the maximum that can be achieved. The ASTM D943 test is also very significant for the long-term stability of an oil. We will review this later in greater detail. The pour points of these oils are not too low (+ 10°F), but this factor is not as important with sintered materials as with wrought materials, as bearings of this type will completely reabsorb the lubricant on shutdown and not leave a heavy oil film which might cause starting difficulties.

The 150 SSU oil is for very-low-torque applications such as flea power motors or units subject to low temperatures. The 300 SSU oil is a good general-purpose lubricant that can cover a broad range of applications. The heavier 750 SSU oil is for heavy-duty applications, oscillating or slow-speed shafts, where extreme boundary lubrication might apply.

Of course, there will be a few applications where a very light oil, or synthetic, might be needed, such as in low-torque applications on aircraft. Conversely, there will always be a few applications, mainly parts, where a heavy-duty fortified oil, such as an SAE 90 Gear Oil, might be needed.

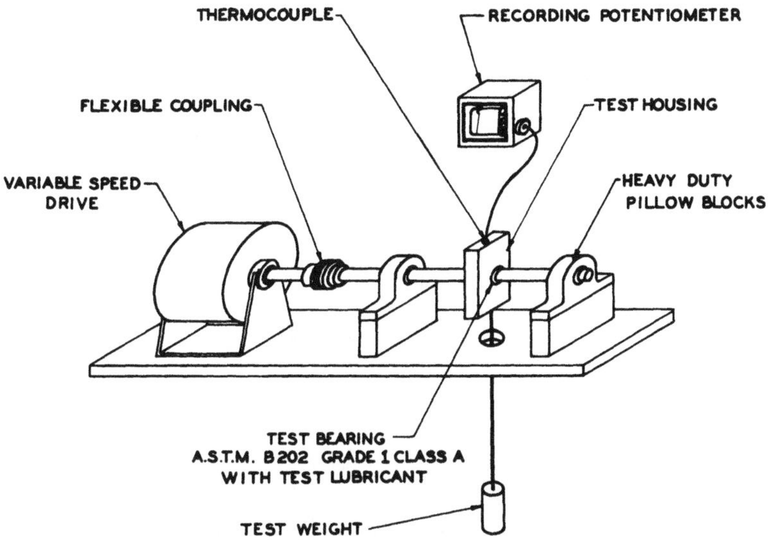

Fig. 1. Test stand.

Qualification Tests

We use two basic qualification tests to evaluate lubricants for use in bearings. The first is a PV = 25,000 eight-hour running test made on a test stand as shown in Fig. 1. The bearing size is a nominal $\frac{3}{4} \times 1 \times 1\frac{3}{8}$, giving an effective bearing area of 1 in.^{2}, and a shaft velocity of 500 fpm. We use a controlled group of bearings, even to having the same permeability, made to ASTM B202 Grade I Class A specification for material and porosity. The bearings are vacuum impregnated with the test lubricant and the oil contents determined. Accurate measurements of the shaft and bearing after installation are made and the temperature during the run is taken with a thermocouple at the outside of the bearing. A recording potentiometer is used which graphically illustrates the break-in, and from these temperature curves we get a good idea of the film strength of the lubricant. After the initial break-in, which might take an hour, the temperature should settle down and show a straight line on the graph. A typical curve taken directly from the potentiometer is shown in Fig. 2. We have found a direct correlation between the coefficient of friction and these temperature values.

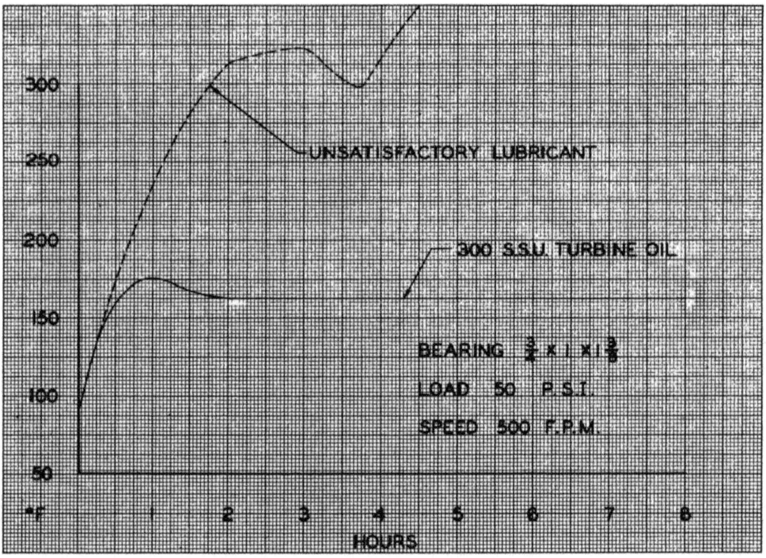

Fig. 2. Bearing temperature during PV = 25,000 eight-hour running test.

At 6.6 oiled specific gravity, without flooded lubrication, the bearing is always operating under thin-film conditions so that the film strength of the lubricant is extremely important. In Fig. 3, we illustrate what is happening between the bearing and shaft. As happens with a convential wrought bearing, the typical wedge formation of hydrodynamic lubrication is built up, but as the wedge nears the load zone, the lubricant disappears into the capillaries and circulates within the bearing wall. This has the advantage of keeping the bearing pores from getting clogged but exerts extra shear on the lubricant and will break down some V. I. improvers and tackiness additives that are susceptible to shear.

Every lubricant must pass this basic test, and practically all petroleum oils do, before we go to the second qualification, which is equally important. This test is a 3000-hour, 100°C test where the lubricant being tested is checked for stability using bearings as a catalyst. It is somewhat similar to ASTM D943, and the results check reasonably close, but we are more interested in the sludging and general condition of the bearing catalyst as well as the test lubricant.

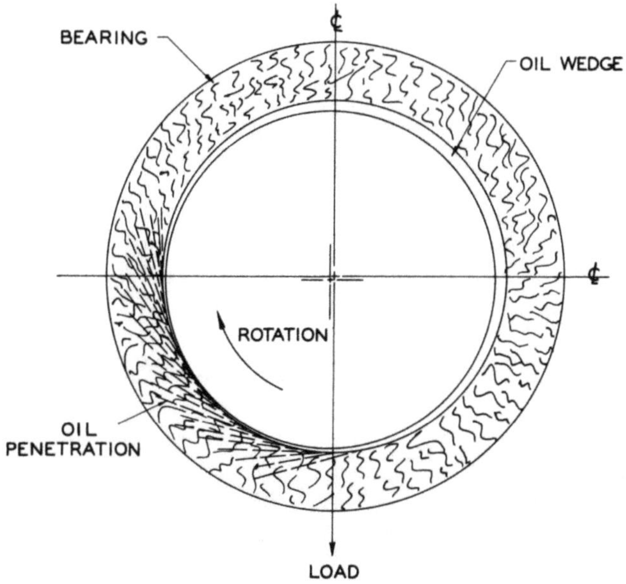

Fig. 3. Lubrication mechanism in a sintered bearing.

This test is made with 110 ml of lubricant in a 150-ml beaker with 75 small bearings, $\frac{3}{16} \times \frac{1}{4} \times \frac{3}{8}$, as the catalyst. The beakers are placed on shelves in an oven held at 100°C ± 2° for 3000 hours. Natural air circulation is used with no excess of oxygen introduced to accelerate the test. Every 1000 hours the shelves are rotated and observations made of the condition of the test lubricant and bearings. Three samples of each lubricant are checked by conventional titration methods to determine acidity. We do not expect a rise in neutralization number of over 0.2 and also use this same value for control of our production vacuum tanks. We discard any oil that exceeds this value, and when it is realized that the government standard is 2.0 for most installations, it can be seen how important we consider this function of the lubricant.

Figure 4 shows the effect of oxidation on the life of an oil and what can be done with inhibitors to increase the life. As can be noted, straight L.V.I. naphthenic oils have very little stability while H.V.I. paraffinic or allied stocks have much greater natural stability but far from what is required. Various antioxidants and/or

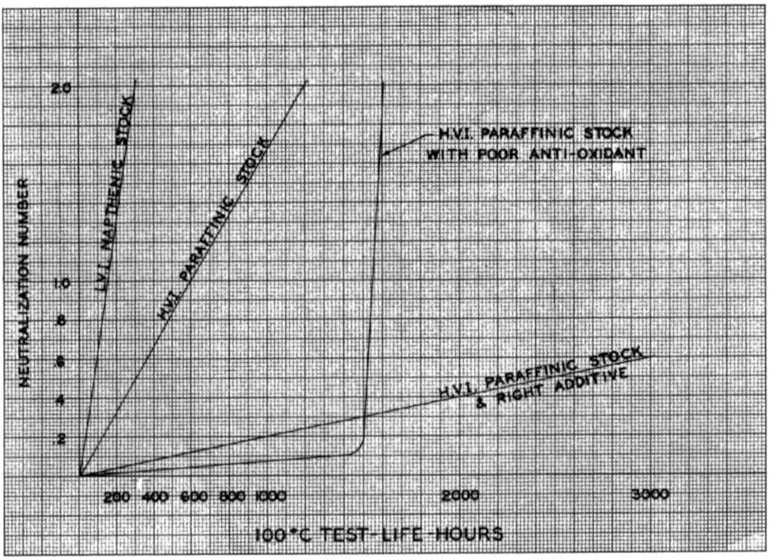

Fig. 4. Effect of oxidation on oil life.

metal deactivators are added in quantities of 0.25 to 1% to increase stability, but the effectiveness of these materials varies with their behavior. Some prevent oxidation almost completely for approximately 1500 hours at 100°C and then completely lose their effectiveness, while others, which we consider more desirable, have slightly higher oxidation rates in the beginning but are effective for a much longer period. Depending strictly on temperature, the life of these lubricants may be 3000 to 4000 hours at the 100°C test temperature but drops fast as the temperatures are raised, until at 130°C the life will be approximately 1000 hours. It is for this reason that considerable effort is being put into different inhibitors and the development of synthetic fluids suitable for high temperatures.

Synthetic Fluids

The development of synthetic fluids as lubricants is progressing in many directions, but the results obtained when they are used with self-lubricating bearings have been disappointing to date except for a rather narrow range of applications. As a group, they have many desirable features such as high viscosity index and oxidation stability, but no one type has all the desired properties. The present status of these various materials is as illustrated in Table 2 with the following comments.

Silicones

This name applies to a rather broad group of lubricants having excellent oxidation stability and a flat viscosity curve but very poor lubriaction properties. They will not pass any of our standard running tests, evidently because of their inability to wet the shaft and their high creep property. They are being used in extremely light-duty timer motors having negligible loads but requiring long life at fairly high temperatures.

Polyglycols

This group of lubricants has excellent film strength giving high load-carrying ability to the bearing and has a V.I. of 140-150. Unfortunately, the oxidation stability is extremely poor due to sublimation of the fluid when in contact with a metal catalyst such as a bearing. The use of convential antioxidants does not appear to correct this defect of an otherwise excellent lubricant.

Silicone–Glycol Blends

This combination appeared to be a natural one to get the high oxidation stability of the silicones and the load-carrying ability of the glycols. Tests indicate that the glycols impart both their good and bad properties to this blend, but while they impart good film strength the high loss due to sublimation is undesirable.

Table 2. Properties of Synthetic Fluids

Fluid	Oxidation stability	V.I.	Lubricity
Silicones	Superior	Superior	Very poor
Polyglycols	Very poor	Excellent	Superior
Silicone-poly-glycol blend	Poor	Superior	Excellent
Diesters	Good	Excellent	Good
Polyol esters	Good	Excellent	Good

Diesters

More diester fluids are used in self-lubricating bearings than any other synthetic. Considerable work has been done, and is still going on, to improve their oxidation stability and uniformity. There is evidence that these materials polymerize to various degrees with time and temperature so that the long-term life at elevated temperatures still has to be proved.

Polyol-Esters

Much work is being done with these fluids and they appear to have much promise for a wide temperature range. Not enough data have been accumulated for firm recommendations as yet.

Fluorocarbons and Chlorocarbons

Neither of these materials appears to have anything to offer as a lubricant for self-lubricating bearings as they are both very unstable in actual installations.

Impregnation Methods

According to ASTM and other engineering and industrial specifications, self-lubricating bearings may be impregnated by either of the following two methods:

1. Reduce the pressure over the specimen immersed in oil held at room temperature to not more than 2 in. Hg for 30 minutes by a suitable evacuating method, and then permit the pressure to increase to atmospheric and the specimen to remain immersed in oil at room temperature and atmospheric pressure for 10 minutes.
2. Immerse the specimen for at least 4 hours in oil (viscosity of approximately 200 sec Saybolt Universal at 100°F), held at a temperature of 180 ± 10°F, and then cool to room temperature by immersion in oil at room temperature.

This is followed by a note that it is preferable to impregnate by the vacuum method especially if the density is over 6.8 g/cm^3. As is shown on Fig. 5, for bearings having a mean specific gravity of 6.6, this note is not strong enough and it is questionable whether we should include immersion impregnation at all in the various specifications except as an emergency method for a customer who

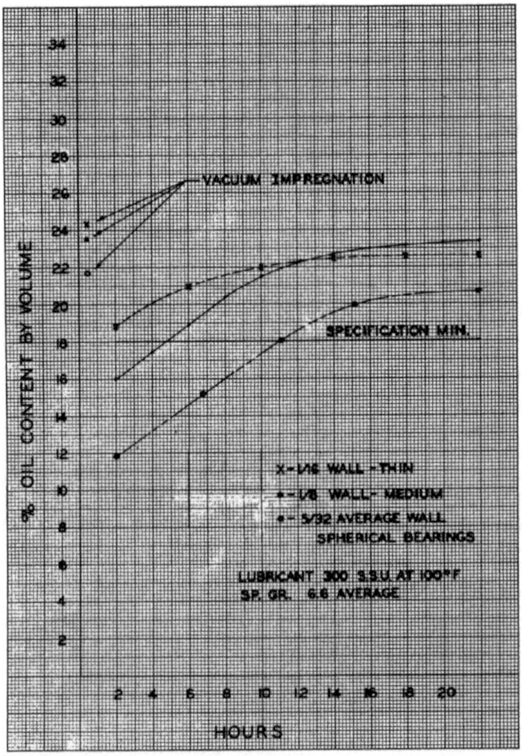

Fig. 5. Immersion versus vacuum impregnation.

wants to change lubricants and is not equipped with a vacuum apparatus. These curves are averages of ten bearings each with $^1/_{16}$-$^3/_{32}$ and $^5/_{32}$ in. average walls using both of the recommended impregnation methods. In the four-hour immersion cycle only the $^1/_{16}$-in. wall bearing met the 18% minimum specification, the $^3/_{32}$-in. wall almost made the minimum, while the heavier-wall bearing was down to 12.6%. Compare these values with the usual 30 minute vacuum cycle where the oil contents are practically up to the full theoretical values. Using the immersion method, it takes approximately 12 hours to approach the vacuum method, and even longer for the heavy-wall bearings.

There is another significant disadvantage to the immersion method, and that is the high impregnation temperature. As we pointed out earlier, in the oxidation test setup temperature is very critical, and oil kept at 180°F for very long periods in the presence of a metal catalyst is going to break down as there will be a high loss of the oxidation inhibitors as well as some of the other additives. The most prevalent inhibitors are the hindered phenols, and these vaporize rapidly at 175°F, and if an oil with some V.I. improvers and tackiness additives is used, they are also adversely affected at this temperature.

The advantages of vacuum impregnation are so great that immersion methods should only be used for emergencies; conversely, there is no advantage, except time, in using both vacuum and high pressure, as is sometimes done, because the vacuum method will fill all interconnected voids.

Supplementary Lubrication

With bearings subject to much higher loads and speeds as well as increased life requirements, the few drops of oil in these bearings must be supplemented with additional lubrication to meet these requirements. In Fig. 6 we show an illustration of typical motor design that incorporates most good design principles.

Fig. 6. Motor permitting supplementary lubrication of bearings.

The bearing is surrounded with wicking, which can be one of many types. The oldest and most prevalent is wool felt. With well-bleached felts there is no longer a problem with animal fats and there are many grades available, but it is advisable to use SAE F5 or F10 as they are an excellent compromise between high oil content and oil retention. Next in popularity is cellulose, either in the form of punched washers or macerated.

The felt and cellulose washers are impregnated with various lubricants, but it is advisable to use the same lubricant as in the bearings. There is wide disagreement about the compatibility of the various additives and base stocks, so it is best to be safe, although our tests have not indicated any significant problem. The macerated cellulose is generally furnished already saturated with different lubricants, generally 85-90% lubricant, with the balance cellulose.

Greases should be avoided where it is expected to feed through the wall. The capillaries are so fine that they are excellent filters and will filter out the various soaps and gels from greases and eventually will become clogged and the bearings will no longer function as self-lubricating bearings. Grease between two bearings in contact with the shaft and the ends of the bearings is satisfactory as the clogging, if it happens, is not so critical as the clearance is also capillary in its action.

As mentioned earlier, we have a problem in finding a lubricant for the higher temperatures, and when we do we must have a means of holding it in suspension for supplementary systems. Felts, being a protein, "cook" and harden at elevated temperatures, and cellulose, being a wood fiber, also dries out rapidly and loses its desirable characteristics. Again, we have to look to the synthetics such as the acetates and polyesters, but the combinations are going to have to be evaluated very carefully for compatibility with the synthetic lubricants or additive-type mineral oils. If the bearing is impregnated with a synthetic, so should the reservoir, and it should preferably be a closed system so that no added lubrication can be applied. If added lubrication is a necessity, it should be provided with the unit and spelled out very plainly in the instructions; otherwise there may be an early failure.

In this same design, provision is made to keep the lubricant in the system. With normal temperatures lubricants don't wear out, but they will pump out, creep, and in general migrate. The object is to control this by the use of slingers, oil return systems, and a wicking system so that there is continuous controlled circulation of the lubricant. If this is done we have a true self-lubricating bearing; if not, we eventually have a dry bearing. But sintered materials require lubrication just as wrought materials, otherwise we would have the same type of seizures and failures.

Chapter 19

Sintered Bearings

Arthur Adler
Easton Metal Powder Company
Easton, Pa.

Sintered bearings can be considered P/M parts with a pore volume of between 17% and 30%. They are easily impregnated with lubricants or other substances, and are ideal for use as bearings with self-lubricating properties. They also are used as compound materials where the sintered alloys are compounded with a base material. The various alloy materials like sintered iron with graphite, copper, or lead additions, and sintered bronze with graphite and/or lead additions, have wide applications. With compound materials copper–tin alloys are widely used as the basic coating material and required properties are obtained by the addition of iron, lead, and graphite; steel plate usually serves as the base or backing material.

Other structural components can be manufactured from sintered materials with high porosity which are not generally classified as bearings. Because of their many applications and varieties, this group of P/M parts is manufactured under several trade names. However, in regard to porosity they are classified in Fig. 1 as a generic group of sintered materials.

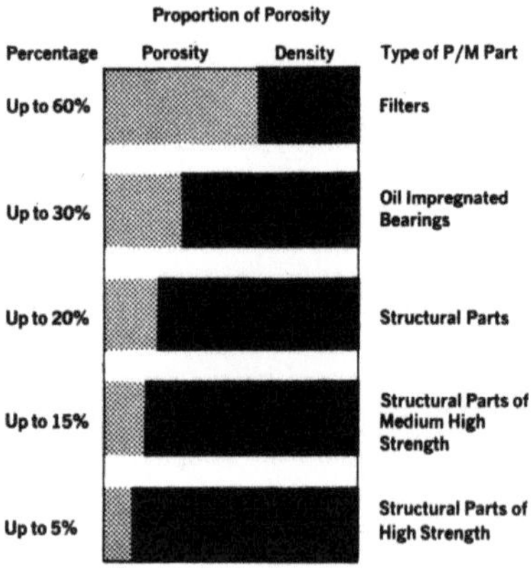

Fig. 1. Powder metallurgy materials classified by percentage of porosity.

Table 1. Physical Properties and Chemical Analysis of Sintered Bearings

Material type	Surface finish	Pore volume, %	Density, g/cc	Tensile strength, psi	Elongation, %	Brinell hardness	Iron, %	Carbon, %	Copper, %	Tin, %	Balance, %
Sintered iron	Smooth	18	5.8-6.3	10,000	3	30	Bal.	0.3	1.0	–	3.0
Sintered iron with copper additions	Smooth	18	5.8-6.3	20,000	3	40	Bal.	0.3	1-5	–	3.0
Sintered iron with copper and carbon additions	Smooth	18	5.8-6.3	30,000	–	70	Bal.	0.4-1.0	1-5	–	3.0
Sintered iron with high copper addition	Smooth	18	5.8-6.3	25,000	2	45	Bal.	0.3	5.0	–	3.0
Sintered iron with carbon and high copper additions	Smooth	18	5.8-6.3	40,000	2	80	Bal.	0.4-1.0	5.0	–	3.0
Sintered bronze	Smooth	18	6.4-7.0	10,000	3	25	–	0.2	Bal.	9.11	3.0

While the chemical and physical properties of sintered bearings vary among manufacturers, Table 1 is a general summation of the properties which can be expected in oil-impregnated bearings. Other properties such as thermal conductivity, electrical conductivity, thermal ductility, torsional strength, modulus of elasticity, etc. are contained in the literature of individual manufacturers.

Bearing Life

The life of a sintered bearing depends largely on the design and method of fabrication. With the oil held only in the pore volume of the bearing, operational life of 3000 to 5000 hours can be reached provided there is no loss due to throw-off or drip. With lubrication reserves such as oil-impregnated felt rings or grease deposit spaces the life of a properly designed sintered bearing can be increased to tens of thousands of hours. The lubrication reserve should be about one or two times the volume of the bearing part. The life of the equipment and its bearings can, through design, be so arranged that maintenance of the bearing points is completely unnecessary.

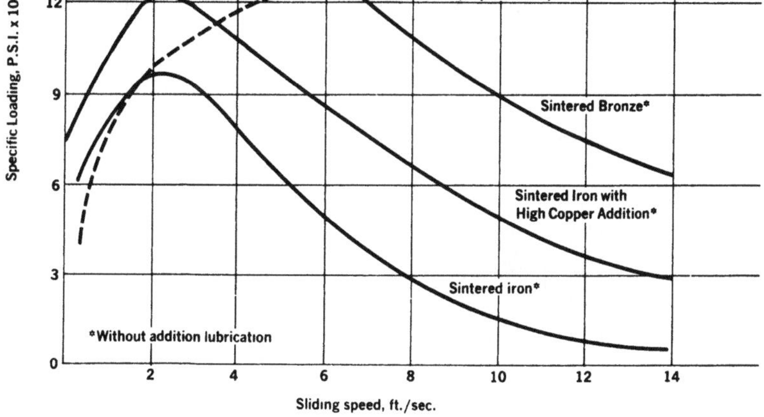

Fig. 2. Radial load capacity of sintered bearings using some of the materials shown in Table 1.

Loading Capacity

Figure 2 is a summary of the radial loading limits of sintered bearings of some of the materials listed in Table 1 in relation to the sliding speed and in comparison with a solid metal bearing. For the four bearings shown in Fig. 2 the dimensions were 20/30 diameter ×20 mm. The impregnated oil viscosity was 15 centistokes at 120°F. The loading was increased in steps until the temperature was no longer constant.

These test values show allowable loading limits under favorable or average working conditions. Sintered bearings will frequently stand higher loading at slower sliding speeds or oscillating movements than solid bearings from wrought metal. The pore volume consisting of numerous oil channels guarantees a sufficient supply of lubricant under unfavorable operating conditions; with a solid bearing, no oil film would be formed. With increasing sliding speed the carrying capacity of the sintered bearings falls off. Practice has shown, however, that even at higher sliding speeds, sufficient life can be achieved. In general, axial loading capacity is about 25% of the radial loading capacity.

Temperature Range

The allowable temperature range for a sintered bearing is largely affected by the quality of the impregnated oil. On one hand the lubricant must be so tough that mechanical oil loss is avoided and on the other hand, the viscosity in the working temperature range must be so low that if possible no additional friction

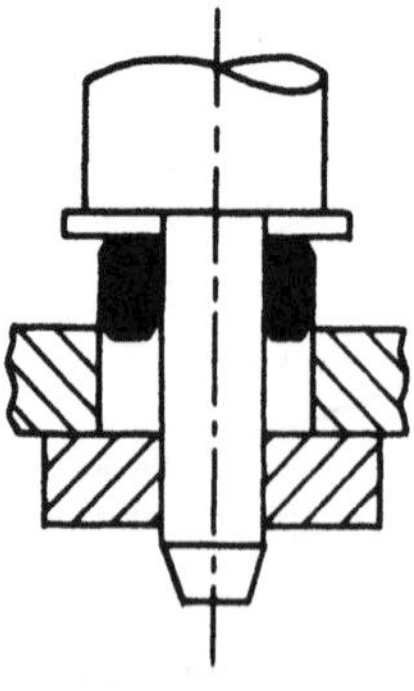

Fig. 3. Cylindrical type of sintered bearing.

loss occurs. By suitable balancing of these points, temperatures of about 100°F to 250°F can be reached with oils having a viscosity of between 12 centistokes and 250 centistokes at 120°F.

Construction

Figure 3 shows the cylindrical-type sintered bearing which must be pressed over a mandrel. The mandrel must have close limits and, if possible, a honed surface finish in order to avoid damaging the bore of the bearing. Should two or more bearings be assembled in line in the housing, they can be pressed into position with the aid of a press ring over the same mandrel as shown in Fig. 4.

Spherical bearings should be assembled with springs which press the bearing firmly, but with a certain amount of give, against the housing. The spring tension should be enough so that when the end of an iron rod of length = (25–30) × d is inserted into the spherical bearing, it will just move from the horizontal with a slight initiating load as shown in Fig. 5. This will insure that the spherical bearing is self-aligning to the direction of the shaft.

If shafts are to be lubricated with oils, it is necessary to feed oil only to the O.D. or end of the bearing. Greases, however,

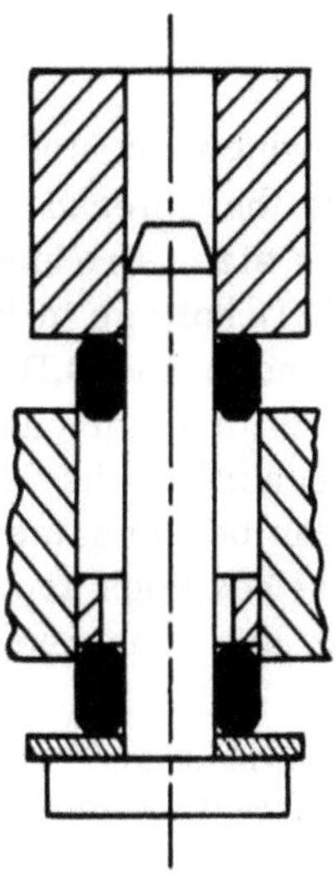

Fig. 4. Two sintered bearings in line. Bearing in position: bore G7, O.D. R6; housing: bore H7.

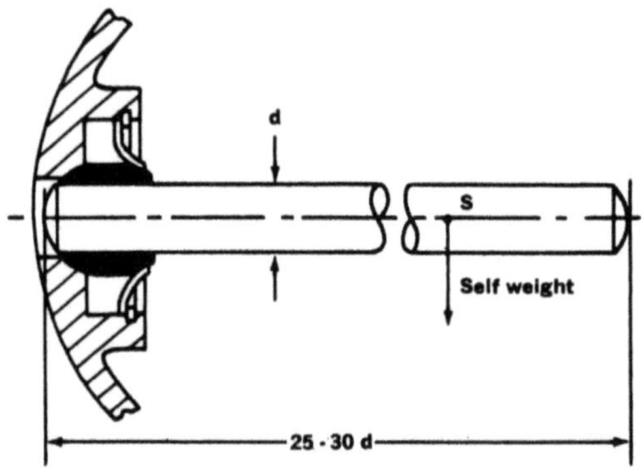

Fig. 5. Spherical sintered bearing.

are fed directly to the shaft through holes and grooves, or exposed ends or sections of the bearing.

Accuracy of Shape

Sintered bearings are generally considered to be precision bearings. By due care in production and quality control, discrepancies in axial squareness, concentricity, and form of bore can be kept to a minimum. Figure 6 shows a number of different types of errors and the way in which they are detected.

Fits

The high precision demands which are today placed on sintered bearings make it preferable that tolerances are given on a standard hole system.* If the bore size is reduced by pressing a cylindrical bearing for a standard H7 hole as in Fig. 4, it should be toleranced in the bore with a G7 and on the O.D. with R6 for a housing with a hole tolerance H7. With spherical bearings no tolerance shift of the bore occurs because of the design. Thus the tolerances of the supplied part can be the same as the working tolerance. For this reason spherical bearings are preferable where small amounts of running play are concerned. For ex-

*Refer to "Specifications for shaft and housing deviation from nominal diameters." The American Anti-Friction Bearing Manufacturers Association, Inc., 60 E. 42nd St., New York, N. Y. 10017.

tremely high demands on noiselessness, and thus accuracy, spherical bearings today are produced in series with bore tolerances up to IT3, preferably H3.

Stock Size

A vast majority of cylindrical and spherical bearings are standarized. Sizes with standard holes should be used wherever possible for new designs. Most manufacturers' size data contain information on usual accuracy, such as length tolerance, axial squareness, and concentricity discrepancies for normal production.

Sintered bearings are mostly supplied impregnated with a lubricant and ready to use. Besides this, semi-finished parts can

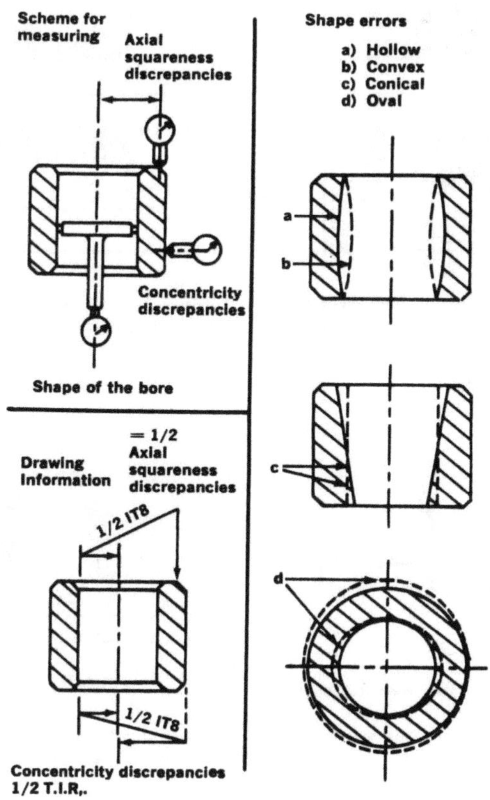

Fig. 6. Types and detection of shape errors.

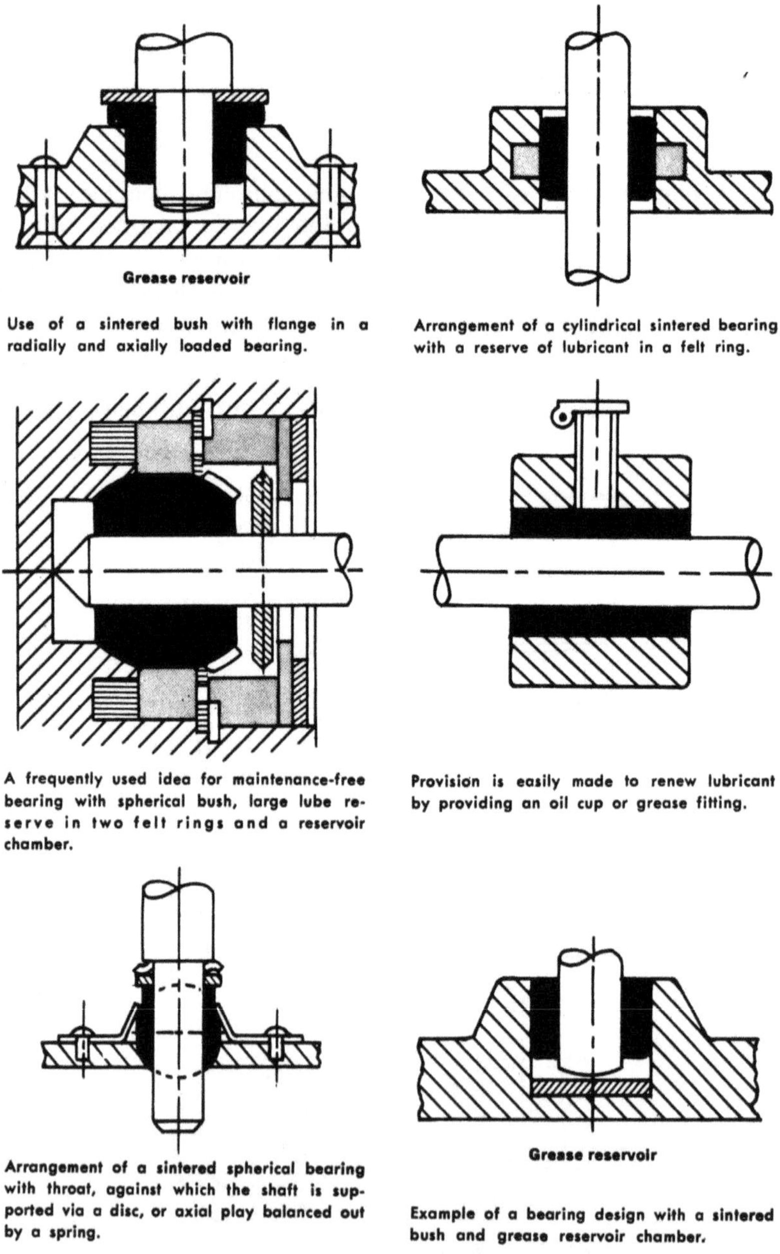

Fig. 7. Six popular bearing assemblies.

be obtained from which the desired shape of bearings can be manufactured by recommended procedures for secondary operations.

Range of Uses

Sintered bearings are to be found in all ranges of technical mechanisms which demand noiseless running, no maintenance, and precision. Because of this, manufacturers offer technical service for answering questions on application and uses. Figure 7 shows some of the more popular types of assemblies.

Chapter 20

Effect of Copper Addition on the Bearing Properties of Sintered Iron-Graphite

A. Domsa

Powder Metallurgy Laboratory
Cluj, Roumania

Mechanical engineering requires that bearing materials have not only high mechanical properties, but also be self-lubricating at elevated temperatures. As liquid lubricants (oil) lose their lubricating ability under such conditions, solid lubricants (e.g., graphite) are preferred. Bearing (antifrictional) materials have been produced by P/M techniques from iron powder and a high portion of graphite and used successfully.

Operational characteristics of iron-graphite P/M bearings depend mainly on the degree of porosity and the structure of the material.

In order to ensure consistent lubricating properties for long-term use, the relative sintered density of compacts made from an iron-graphite mixture should not be very high. The residual porosity should be at least 20% to retain a sufficient amount of lubricating oil. If the particle size of the powder is constant, then the degree of porosity of the finished piece depends primarily on the compacting pressure, conditions of sintering, and the amount of added graphite (Fig. 1).

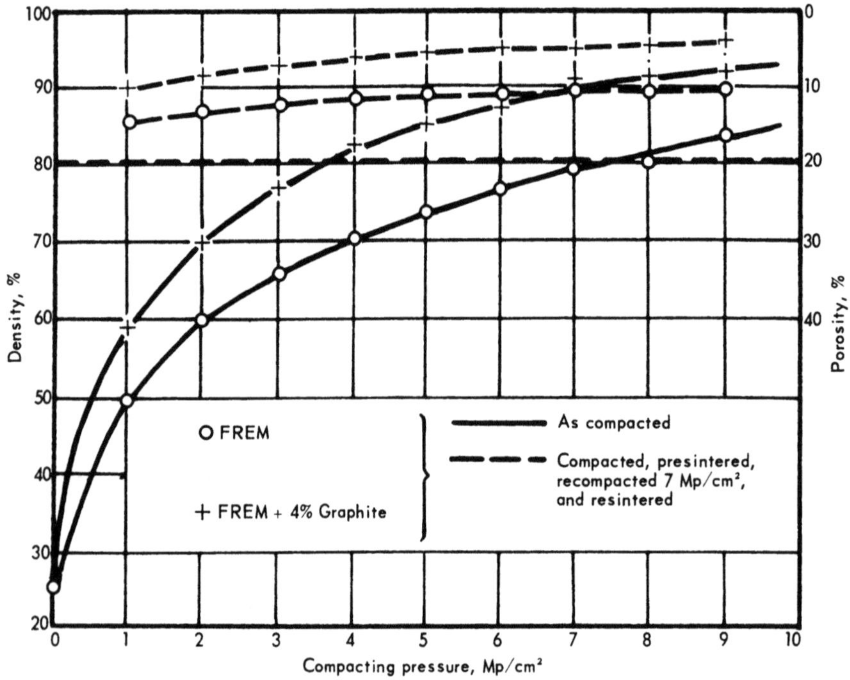

Fig. 1. Density as a function of compacting pressure and sintering conditions.

When the grade FREM [1,2,3] of iron powder (see Table 1) was used, the desired minimum porosity of 20% was realized only by compacting at loads above 7 Mp/cm^2. Additions of graphite considerably increase the density at the same compacting pressure. Specimens with an addition of 4% graphite displayed porosity below 20% even at a 4 Mp/cm^2 compacting pressure (see Fig. 1).

Another difficulty connected with the manufacture of iron-graphite P/M products is the appearance of free cementite under

Table 1

Powder	Fe, %	C, %	Max. O, %	Apparent density, g/cm^3	Tap density, g/cm^3	Grain size, mm
FREM (by reducing of oxides)	98	< 0.09	1.5	1.86	2.5-3.0	<0.15
Graphite	—	<96	—	0.35	—	90% < 0.06

normal sintering conditions. For the purpose of preventing the formation of free cementite in materials with higher graphite contents, the sintering time is often cut short or a multistage method of sintering is adopted [4, 5]. Even if these measures are taken, the content of graphite may not exceed 3%.

With the aim of obstructing the diffusion of carbon in iron and thus improving mechanical properties, the addition of various alloying elements has been investigated. In this respect, copper has proved most effective. According to the binary equilibrium diagram (Fig. 2), copper is soluble in α-iron to a limited extent. The solubility increases with rising temperature and reaches about 8% in the γ-region near the melting point of copper (1083°C). The alloying of ferrite by copper [6] causes a pronounced subsequent suppression of diffusion of carbon in γ-iron and thus pre-

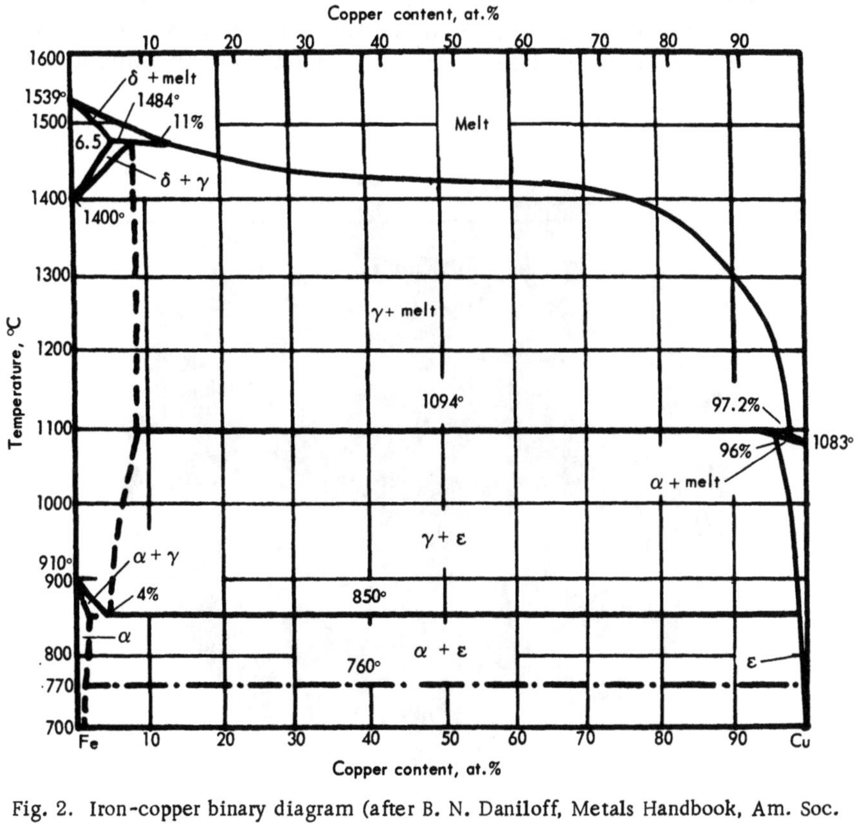

Fig. 2. Iron-copper binary diagram (after B. N. Daniloff, Metals Handbook, Am. Soc. Met., Cleveland, 1948, p. 1196).

Table 2

	Graphite, %	Compacting pressure, Mp/cm²	Sintering	
			°C	min
Compacts	0-4	1-9	Presintering 800	60
		Coining 2, 4, and 7	Final sintering 1050	10-25
Bushes	0-4	4, 6	1050	30-60

vents the formation of cementite. The decreasing solubility of copper upon cooling can cause some increase in strength due to precipitation hardening [7, 8, 9].

The behavior of copper in iron-copper-graphite P/M alloys is rather complicated and not widely understood. The sintering phenomena and the properties of the material depend (1) on the sintering time and temperature, (2) on the percentages of the powder mixture components, and (3) on the subsequent heat treatment [10].

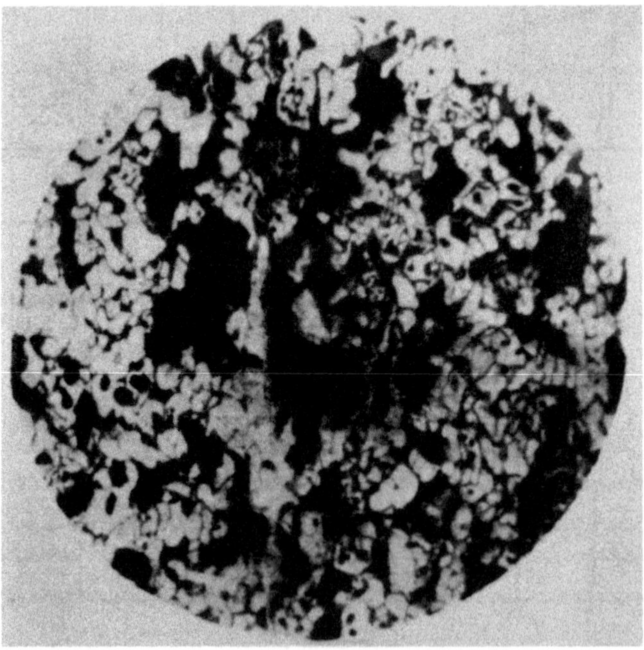

Fig. 3. Micrograph of Fe + 4% graphite + 7% Cu, presintered at 800°C (300 ×).

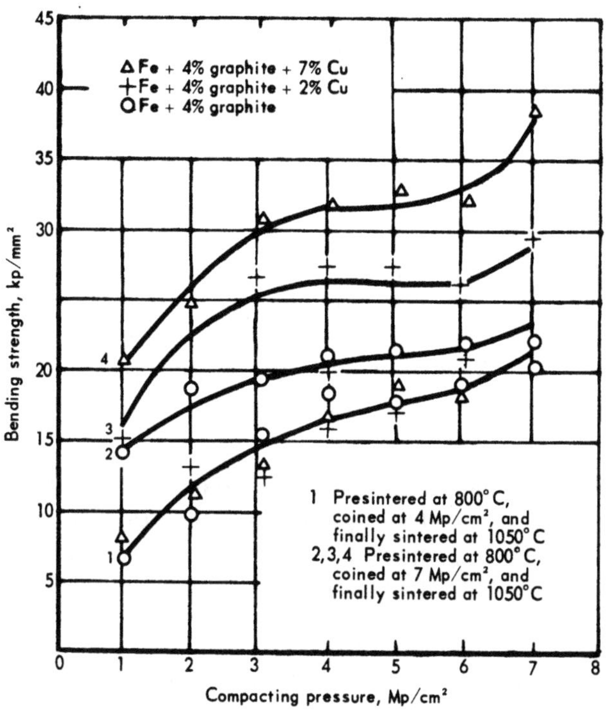

Fig. 4. Bending strength as a function of compacting pressure and sintering conditions.

The following basic materials were used for the present investigation: iron powder grade FREM with a low carbon content [3]; copper, atomized by a special method from the liquid state; graphite (see Table 1). The measured technological parameters are given in Table 2.

The diffusion of copper into the iron lattice takes place during a presintering period, due to an increased solubility of copper in α-iron at 800°C. This will cause a decrease in the diffusivity of carbon in iron and, consequently, suppress the formation of free cementite on cooling. Under these circumstances, the presintering at a temperature just below the Ac_3 point has a decisive effect, as copper can sufficiently diffuse in α-iron, while graphite undergoes practically no change. The structure in Fig. 3 shows crystals of ferrite into which a certain amount of copper has diffused, together with free copper, graphite, and pores. In the last stage of sintering, when the γ-region has been reached, the γ-iron crystals contain some dissolved copper from the preceding presintering and therefore are able to accept less carbon than is shown by the iron-carbon equilibrium diagram. Because of this, sintered materials with high graphite content (4 to 12%) can be produced that are free of cementite, but possess increased mechanical properties (Fig. 4).

The final structure is strongly affected by the sintering time and temperature. The amount of ferrite decreases with increasing sintering time and, after 30 minutes, it takes the form of a network (Fig. 5). The absence of free cementite and the ferritic nature of this network were proved by a sodium picrate etch and microhardness tests. At the end of sintering no free copper was observed.

This result indicates that many possibilities exist in the preparation of self-lubricating P/M materials with solid or semisolid lubricants. Suitable properties for any given requirement can also be attained by varying the conditions of manufacture. In the case of an iron-base material containing 7% copper and 4% graphite, an almost constant coefficient of friction of 0.1 was achieved at a sliding velocity of 0.94 m/sec in the loading range from 10 to 45 kp/cm^2.

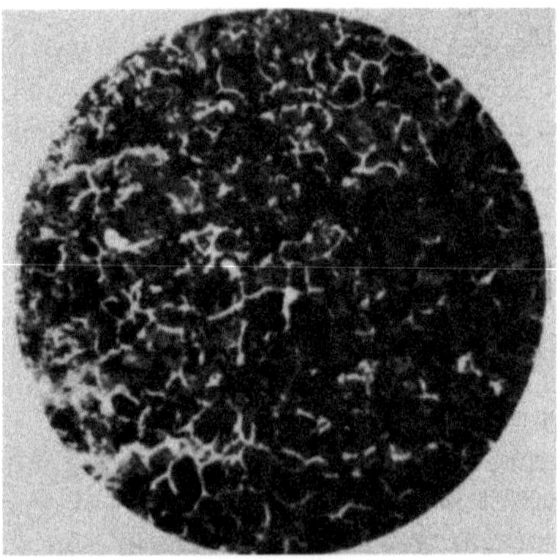

Fig. 5. Micrograph of Fe + 4% graphite + 7% Cu; Presintered at 800°C, sintered at 1050°C for 30 minutes (253×).

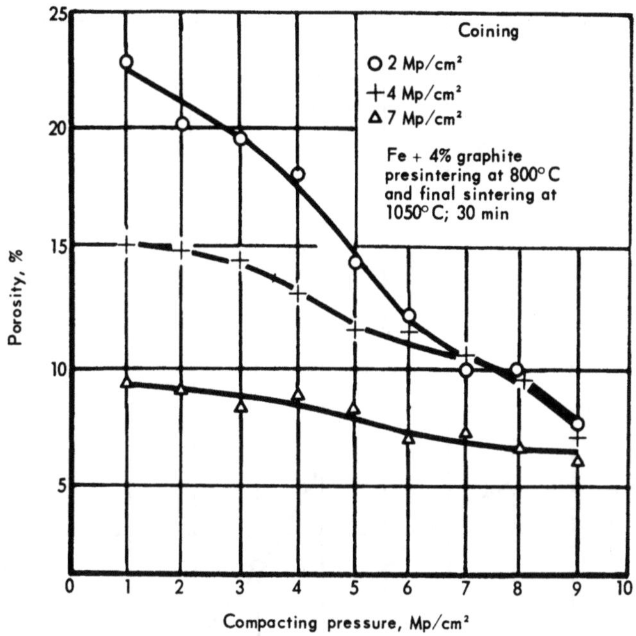

Fig. 6. Porosity as a function of compacting pressure and sintering conditions.

The relationship of porosity and compacting pressure of iron-graphite is not significantly altered by copper additions. Figure 6 shows the porosity as a function of compacting pressure for a copper-free material with 4% graphite coined at various pressures. The corresponding curves for a material containing 7% copper and the same amount of graphite are given in Fig. 7 for comparison. As expected, porosity decreases sharply with increasing coining pressure.

Conclusions

The effect of copper on the diffusion of carbon has been investigated for the purpose of improving the mechanical and technological properties of sintered iron-graphite materials. It has been proved that copper strongly reduces the diffusion of carbon in iron and thus improves the metallic matrix of these materials. The self-lubricating ability of these materials can be achieved by the following means:

1. Under mild conditions; by impregnation with oil under vacuum, without additional lubrication.

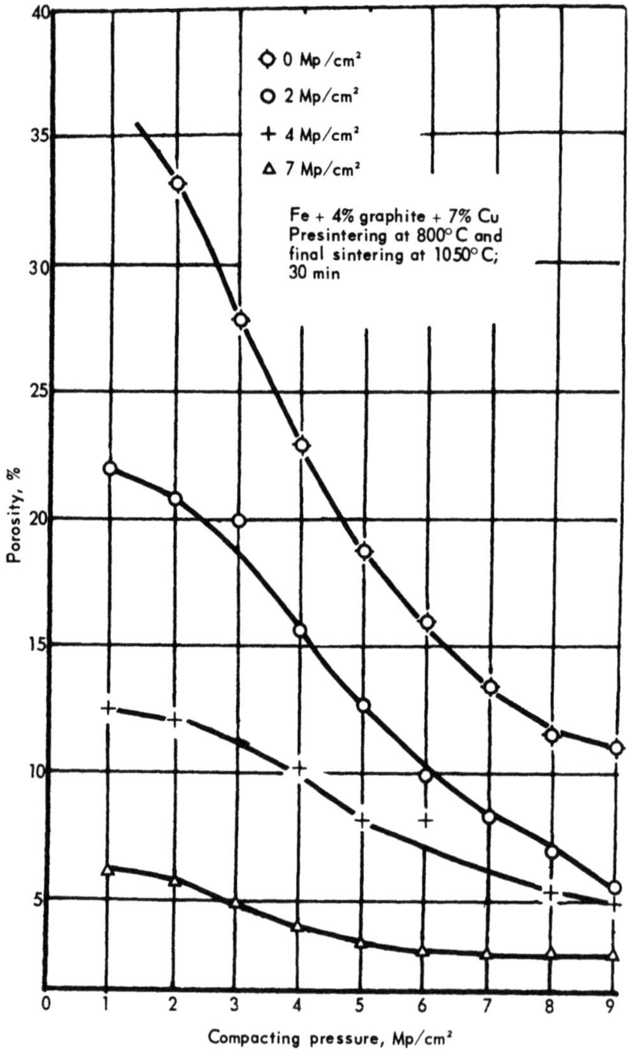

Fig. 7. Porosity as a function of compacting pressure and sintering conditions.

2. Under intermediate conditions (velocity above 1.5 m/sec): either by oil impregnation + external oil addition, or by oil impregnation and the presence of free graphite in the bearing material.

3. Under very heavy and extraordinary conditions (high loading and high temperatures): only by the presence of free graphite in the bearing material.

Summary

It is difficult to give generally valid data for the preparation of sintered materials with sliding properties defined in advance, mainly because of the different characteristics of iron powders obtained from various sources. Iron-graphite materials possessing higher amounts of graphite are liable to the formation of free cementite under normal conditions of sintering, which deteriorates the bearing properties. In the research performed the effect of copper additions has been investigated. The experiments proved the retarding effect of copper on the diffusion of carbon in iron at elevated temperatures. This results in the suppression of cementite formation and, consequently, in the improvement of the bearing properties.

References

1. Domsa, A., and Müller, G., Studii si cercetari de metalurgie, Academia R. P.R., Nos. 1 and 2, 1957.
2. Domsa, A., and Colan, H., Studii si cercetari de metalurgie, Academia R.P.R., Nos. 1 and 2, 1957.
3. Domsa, A., and Palfalvi, A., Studii si cercetari di metalurgie, Academia R.P.R., No. 1, 1959.
4. Bebnev, P. I., "Formation of Structure in Iron-Graphite during Sintering," Sbornik poroshkovaya metalurgiya, Yaroslavl 1956.
5. Bebnev, P. I., "Porous Iron-Based Bearings with Improved Antifrictional and Mechanical Properties," Sbornik TsNIITMas, No. 56, 1953.
6. Revue de Metallurgie, No. 11, p. 879-888, 1960.
7. Arch. Eisenhuttenw. No. 12, p. 785-792, 1958.
8. Arch. Eisenhuttenw. No. 2, p. 123-127, 1961.
9. Stahl u. Eisen Vol. 79, p. 1187-1201, 1959.
10. Metallovedenie, No. 12, p. 42-43, 1960.

Chapter 21

Effect of Sizing Allowance on the Surface Quality of Iron-Graphite Bushings

D. Z. Yurchenko

Institute of Materials Science
Academy of Sciences of the Ukrainian SSR
Kiev, USSR

Information was previously published [1, 2] concerning the effect of the sizing allowance on the surface finish, work hardening, and compaction of porous iron during sizing. The results of these investigations were used for the selection of optimal values of the sizing allowances.

The results of similar investigations with respect to iron-graphite bearings have not yet appeared in the literature, although it would be important to establish the effect of the structural factor on the kinetics of the transformation of the physicochemical state in the surface layers of iron-graphite materials. The Institute of Materials Science therefore studied the pattern of change in the degree of compaction and hardening of the material in the surface layers of iron-graphite porous specimens in relation to the sizing parameters. In the investigation we used cylindrical specimens 10mm in diameter × 10 mm with a porosity of 10-30% pressed from a mixture of TZhlM iron powder and 3% kLZ flaky graphite. The samples were sintered at 1050°C for 2 hr in a dry hydrogen atmosphere. The structure of the material was pearlitic-cementitic.

The surface layers on the ends of the specimens, which were similar to the surfaces of sized bushings, were obtained in conformity with separate sizing by pushing the specimen through a press-aperture, and in conformity with combined sizing by swaging in a steel casing under conditions of triaxial compression. For a constant starting state the assembly specimen-casing was coupled at the outset by a fine fit. The sizing allowance of the specimens for both types of sizing varied from 0.1 to 0.8 mm.

The degree of hardening was estimated by means of x-ray diffraction analysis (the change of line width or magnitude of distortions of the second kind) and microhardness measurements. The sized surfaces of the specimens were subjected to sequential x-raying with gradual layer-by-layer removal of the metal by electrolytic tippling. As a result of this we established the character of the distribution and depth of penetration of work hardening. A URS-50I ionization device and Fe radiation were used; a sintered iron-graphite specimen with pronounced separation of the doublets in the (220) plane served as the standard. The x-ray patterns were read by the method of photographing the specimen with the standard [3].

Plots of the variation of (220) line width characterizing the total changes of the submicrostructure and distortion of the crystal lattice with respect to depth of the work-hardened layer are shown in Fig. 1 for the separate method of sizing. We see from the figure that the greatest broadening of the (220) lines occurs at the surface; with increasing distance from the surface the (220)

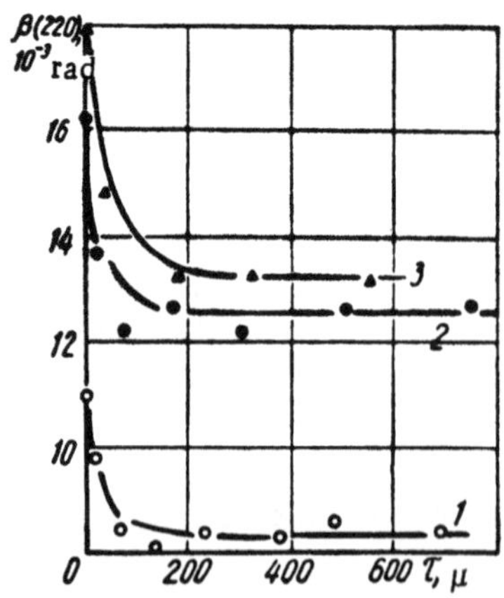

Fig. 1. Change of line width (220) with respect to depth of specimens with P = 21% sized by the separate method with an allowance Δ of (1) 0.09 mm; (2) 0.33 mm; (3) 0.55 mm.

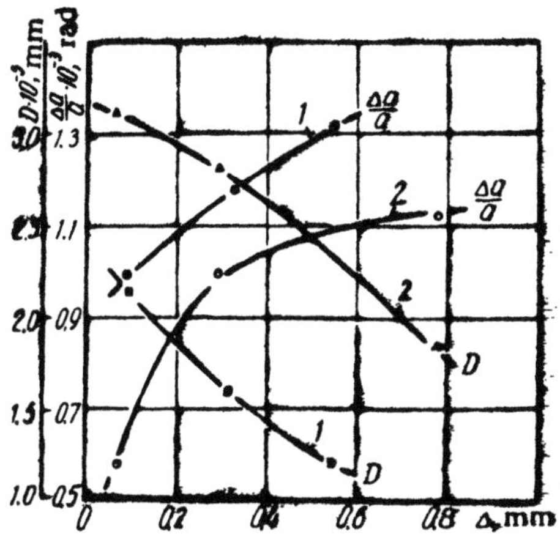

Fig. 2. Effect of sizing allowance on magnitude of distortion of the crystal lattice and size of blocks of specimens with a porosity of 21% (1) and 31% (2) (separate method of sizing).

line width decreases, becoming stabilized at a certain point, however without reaching the line width of the standard. A comparison of the experimental data obtained in separate sizings of specimens of different porosity showed that with an increase in the allowance and decrease in porosity the degree of change of the fine structure expressed by the (220) line width increases and reaches a maximal value of 19.4×10^{-3} rad at an allowance of 0.62 mm for material with 12% porosity. A minimal value of the line width (9.8×10^{-3} rad) was established in x-raying a specimen with 31% porosity sized with an allowance of 0.07 mm.

In combined sizing we observed qualitatively similar changes of (220) line width in relation to porosity and sizing allowance. However, a greater degree of change of the submicrostructure is inherent in the separate method of sizing. The maximal (220) line width of 15.6×10^{-3} rad was recorded when sizing specimens with 12% porosity and an allowance of 0.52 mm.

The results of measuring the magnitude of distortion of the crystal lattice and size of the mosaic blocks of specimens sized with various allowances are shown in Fig. 2 for separate sizing. We see from the data that with an increase in the density of the specimens and sizing allowance there is a quite intense crushing

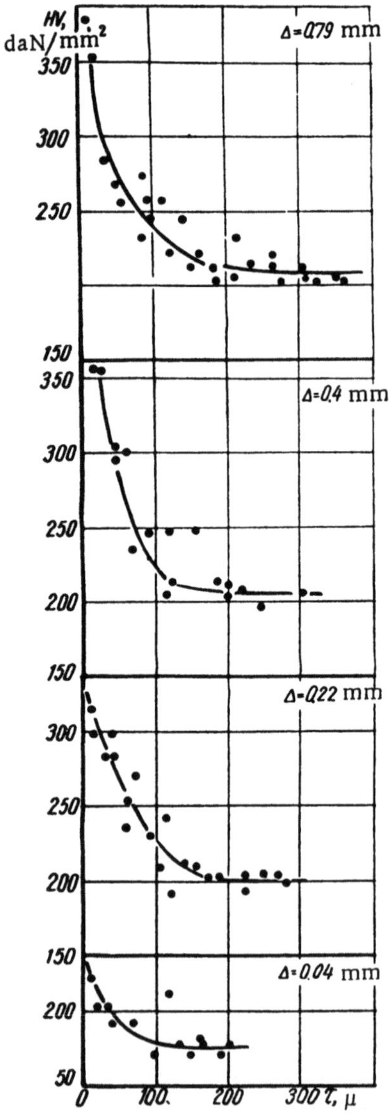

Fig. 3. Change of microhardness with depth of specimens with P = 10% sized by the separate method with various allowances.

of the blocks of the mosaic structure and an increase in the magnitude of distortions of the second kind.

In the case of combined sizing the magnitude of the distortion of the crystal lattice is lower, and the mosaic blocks are larger, other conditions being equal. Whereas in separate sizing of specimens with a porosity of 21% $\Delta a/a = 1.2 \times 10^{-3}$ rad and

D = 1.4 × 10^{-5} cm, in the case of combined sizing $\Delta a/a$ = 0.95 × 10^{-3} rad and D = 2.1 × 10^{-5} cm at an allowance of 0.4 mm. An estimate of the hardening of the surface layers by the microhardness method confirms the data obtained by the x-ray diffraction method. The microhardness was measured by a PMT-3 hardness gauge on an oblique microsection which was ferruginized to prevent its edges from curling. The load on the indentor was 20 g. Figure 3 shows the variation in microhardness as a function of the sizing allowance with respect to the depth of the hardened layer. Just as the (220) line width, the microhardness is greatest near the surface, then drops quite rapidly and, finally, at a certain depth becomes stabilized without reaching the microhardness of iron-graphite in the initial state (115 daN/mm²).

For both cases of sizing under consideration the increase in microhardness with an increase in allowance and density is regular. However, the degree of work hardening for the same values of the allowance and porosity is higher for the separate method (see Table 1). As regards the depth of penetration of intense hardening, the experiments showed that it increases with an increase in the sizing allowance and porosity and is within 80–200μ.

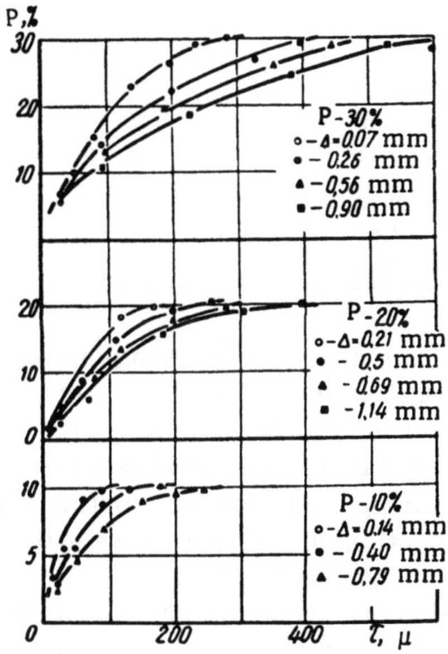

Fig. 4. Change of porosity with depth of specimens sized by the separate method and with different allowances.

Table 1. Effect of Porosity and Allowance on Microhardness of Sized Iron-Graphite

Porosity of specimen, %	Sizing allowance, mm	Method of sizing	Microhardness, daN/mm² at a depth of 20 μ	Microhardness, daN/mm² at the moment of stabilization
10	0.79	Separate	375	200-207
	0.4		350	
	0.22		305	
	0.07		212	
20	0.74	"	275	175-170
	0.42		212	
	0.22		200	
30	1.06	"	215	120
	0.63		185	
	0.30		170	
	0.08		135	
10	0.70	Combined	275	—
	0.51		240	
	0.26		215	170-175
	0.09		205	
20	0.76	"	190	145-150
	0.48		180	
	0.15		160	
31	0.73	"	175	115
	0.57		152	
	0.20		145	

During sizing, an appreciable change in porosity occurs, in addition to hardening, in the surface layers of porous iron-graphite specimens as a result of plastic deformation of the material. Compaction of the surface layers plays a major role in the formation of the surface which during operation is a mating surface. Therefore, a study of the compaction of surface layers under different sizing conditions is of considerable interest.

The compaction of porous material with respect to the depth of the surface layer was studied by means of the NIN-8 microscope by the planimetric method [4]. From the plastically deformed end surface of the specimens we compared a microsection, recording the thickness of the layer removed when preparing the thin section. To avoid plugging of the pores on the surface, the microsection was electrolytically pickled for 30 sec under experimentally predetermined conditions which precluded corrosion of the pores. The metal layers were removed by abrasive paper.

Some of the results of these experiments are shown in Fig. 4, from which we see that the porosity of the surface layer of the specimens with a different initial porosity changes substantially in the case of separate sizing. Thus, the surface porosity of specimens at a depth of 15-25μ is minimal and (depending on the magnitude of the initial porosity and sizing allowance) amounts to 2.5%-7%. With increasing distance from the surface the porosity in the specimens increases, and then becomes stabilized, reaching the value of the initial porosity.

The depth of the compacted layer depends on the initial porosity and the sizing allowance. With an increase of these quantities, as we see from Fig. 4, the depth of the compacted layer increases and amounts to 60-70μ at $P = 10\%$, $\Delta = 0.14$ mm and 500-550 μ at $P = 30\%$, $\Delta = 0.90$ mm. The qualitative change of density of the material in the surface layer of the specimens sized by the combined method differs little from the aforementioned. However, in the latter case an appreciable depth of the compacted layer is observed under equal conditions. Furthermore, in the case of the combined method of sizing a lower gradient of surface layer density is observed than in the case of separate sizing, as indicated by the slope of the curves.

Conclusions

1. X-ray diffraction and other modern methods of investigation established that during sizing of iron-graphite porous materials, just as for iron, the surface layers are appreciably hardened. The maximal hardening in sizing of low-porous (10%) materials with a large (0.6 mm) allowance reaches a factor of 3-3.5, i.e., the microhardness increased from 110 to 350 daN/mm^2, and line broadening increases from 6-7 $\times$ 10^{-3} to 19 $\times$ 10^{-3} rad.

2. It was established that during sizing the outermost layers of the porous materials are compacted in addition to being hardened and reach the density of a compact material. With an increase in the initial porosity and sizing allowance the depth of the compacted

layer increases, whereas the change in surface layer porosity diminishes.

3. A comparison of the characteristics of the surface layers obtained by the separate and combined methods of sizing indicates that the latter has advantages. It makes it possible to produce in one pass an article with a strongly compacted, hardened outer surface forming a strong, immobile compound and a moderately compacted inner surface having high antifriction properties.

References

1. A. Ya. Artamonov, Poroshkovaya Metallurgiya, 1962.
2. A. Ya. Artamonov and D. Z. Yurchenko, Poroshkovaya Metallurgiya, No. 6, 1964.
3. L. I. Lysak, in: Problems of Metal Physics and Physical Metallurgy, Izd. AN UkrSSR, Kiev, 1955.
4. M. E. Blanter, Method of Investigating Metals and Analysis of Experimental Data, Metallurgizdat, Moscow, 1952.

Chapter 22

Production and Properties of a New Porous Bearing

H. Youssef and M. Eudier

Metafram
Beauchamp (S. & O.), France

Introduction

When a shaft turns in a self-lubricating bearing, the oil circulates around the shaft. If a force is applied to the shaft, the clearance between the shaft and the bearing is reduced in the direction of the stress. Therefore the oil is forced into an increasingly small section. As a result the oil pressure increases in the zone of lesser clearance. This pressure prevents any contact between the shaft and the bearing. A depression occurs in the opposing bearing zone (Fig. 1).

We have determined the pressure distribution for a bronze self-lubricating bearing experimentally. Figure 2 indicates this pressure with a polar representation, while Fig. 3 represents it with Cartesian coordinates. The X-axis gives the angle θ, shown in Fig. 1, which the vector passing through the measurement point makes with the downward vertical line, the load being upwards.

The sum of the upper and lower areas is equal to the total force applied to the shaft, multiplied by a coefficient. The pressure P is equal to the total load C (31 lb) divided by the product

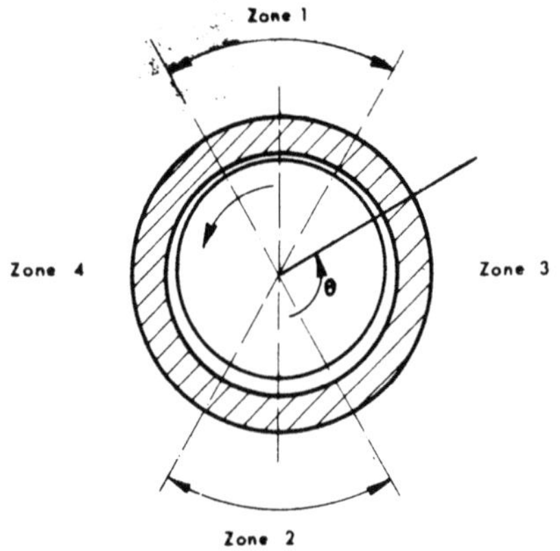

Fig. 1. Section of a shaft inside of bearing. Load is appiled upwards to the shaft

of the diameter D (1 in.) and the bearing length l ($\frac{3}{4}$ in.). The sliding speed V equals the product of the number of runs per minute N and πD. The value of the PV factor is then 15,000 psi/ft-min, and the maximum pressure measured is equal to a proximately 6.5 times the average pressure (41 lb/in.^{2}).

Thermal Balance – Friction Coefficient

Even when there is no metallic contact at all, the oil film is heated by the sliding action of oil molecules in relation to one another. The temperature of the shaft, of the bearing, and of all the parts in contact around them increases. In a continuous operation,

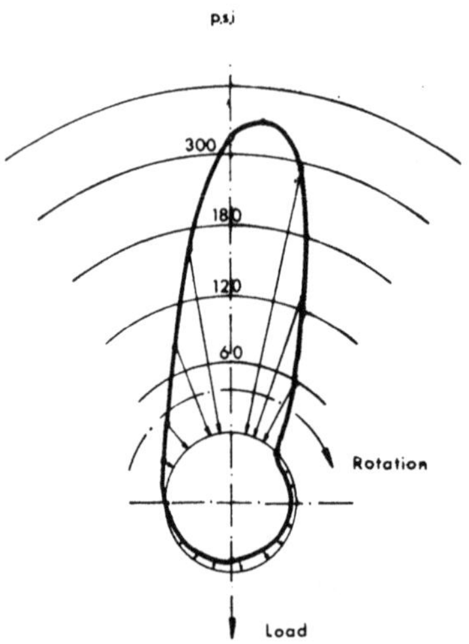

Fig. 2. Pressure distribution of the oil film. Dotted circle is a mean load. Plain cicle is the line of zero pressure.

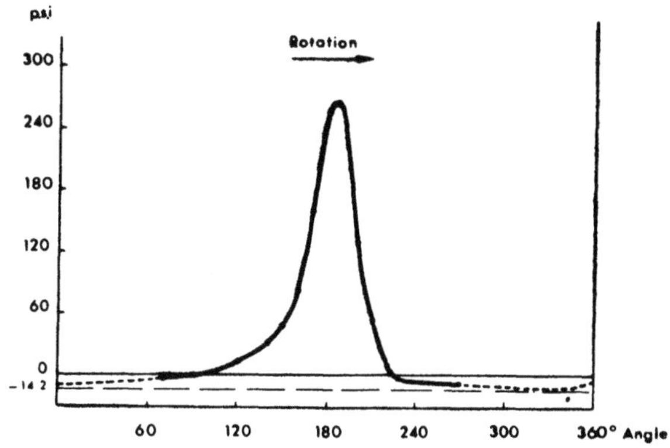

Fig. 3. Pressure in the oil film as a function of the angle θ of Fig. 1.

an equilibrium is reached between the heat created and the heat transmitted to the surrounding atmosphere. This equilibrium produces a definite oil temperature. Furthermore, this temperature depends particularly on the cooling surfaces of the metal parts surrounding the bearing. It also explains why the results obtained from different assemblies are not similar. A better cooling is obtained with an aluminum casing than with a cast iron casing of the same size. Likewise, aluminum bearings drain heat better than bronze bearings. They would be of much interest if they did not have the disadvantage of wearing very quickly when the oil film is not perfect.

On the other hand, the coefficient of friction varies according to the load (Fig. 4a). As a certain coupling is necessary to keep the movement going when the load is zero the coefficient of friction is infinite, which explains the slope of the experimental curve. The load is indicated in the abscissa. Figure 4b represents the bearing temperature as a function of the load for the same tests.

As a practical consequence of the data given in this graph, it seems beneficial to use as short a bearing as possible within certain limits. To illustrate, the released heat is proportional to

$$\frac{C}{S}f = Plf \tag{1}$$

where l is the bearing length, C is the total load, and S is the pro-

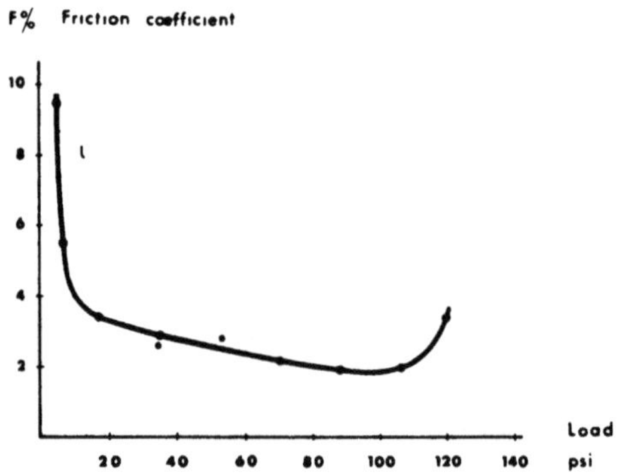

Fig. 4a. Friction coefficient as a function of the load (3000 rpm, shaft diameter 1 in, length $^3/_4$ in.).

jected surface $D \times l$. When using twice as short a bearing, the heat is equal to

$$\frac{C}{S}f = 2Pf'\tfrac{1}{2} = Plf' \tag{2}$$

Referring to the curve in Fig. 4a, one can see that f' is less than f, since f decreases with the load. Consequently, the friction produced by a short bearing is less than the friction produced by a longer one, which results in a lower oil temperature. The lesser length limit is reached at the point where the oil film is discontinued when the load becomes too great.

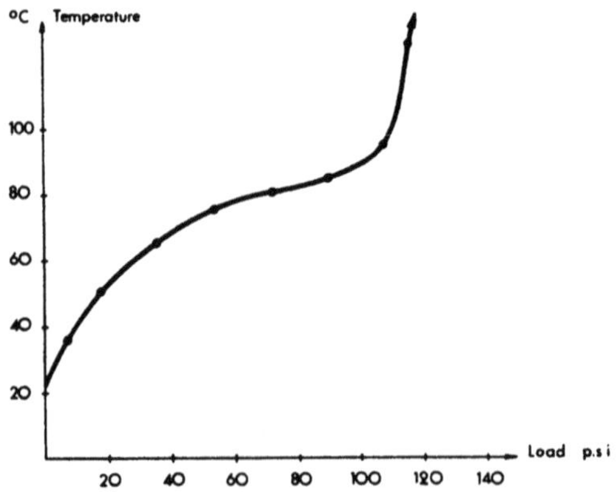

Fig. 4b. Equilibrium temperature as a function of the load.

It has been shown so far that reduction of the length of a bearing may improve its operating conditions. If this is true, the dissipated heat must be the same in both cases, or at least not reduced too much by the reduction in length. This result has been confirmed in practical applications in which the load per unit length was to low.

Oil Circulation

Most of the oil circulates around the shaft. However, with the effect of the pressure in the zone of reduced clearance, part of the oil is expelled laterally, forming an oil ring at each end of the bearing, and another part is forced through the pores of the bearing. These two leakages reduce the pressure which should be reached by the oil flow. The leakages may cause rupture of the oil film, which is one of the causes for bearing failure. To compensate for the loss of oil, the depression which takes place in the sector of large clearance causes a suction effect on (1) part of the oil rings at both ends of the bearing and (2) part of the oil within the bearing. Lateral leakages occur under a high pressure effect (260 psi in this investigation), while the oil recovery takes place with a maximum depression of the atmospheric pressure.

Fortunately, lateral openings in the section under pressure (zone 1) are smaller than lateral openings in the section under depression (zone 2), as shown in Fig. 1.

Returning to the first example, it is known that in order that the oil pumping be equal to the oil losses, the ratio of lateral clearances must range around 3. Moreover, the minimum thickness of an oil film (working under the above conditions) ranges about 5 μ, which results in a total minimum clearance of

$$5 + (3 \times 5)\,\mu = 20\,\mu \text{ (8/10,000 in.)}$$

This indicates that a minimum clearance does exist which depends on several variables, and one of its limitations is due to lateral leakages. A smaller clearance would result in a reduction of the maximum bearing load, due to the rupture of the oil film which escapes too quickly. The functional clearance values recommended by Amplex in particular appeared to be excellent and could be reestablished theoretically with a few assumptions. On the other hand, the oil rings as well as the oil of the film are taken quantitatively from the initial oil content of the bearing. As

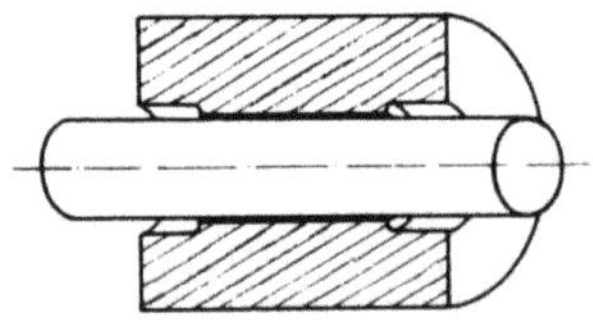

Fig. 5. Grooves made at each end of a bearing to diminish the oil losses.

a result, a certain portion of the bearing becomes dry quickly, and the depression causes some air to enter the oil film. This air then bubbles out of the flanges of the bearing in the pressure zone. In addition, the point where air escapes varies in accordance with the load as the pressure of escaping air approaches more and more closely the horizontal diameter when the load is increased.

With regard to the oil forced through the bearing pores and then pumped into zone 2 under depression, its balance always shows a deficit. Consequently, a very long bearing with relatively small lateral leakages should not work. This has been confirmed in practice. Fortunately, however, another phenomenon improves this condition. When the oil film is broken, the shaft comes into contact with the bearing and polishes it. The pores in the pressure zone are then partially closed up and consequently the oil flow forced through is reduced. The bearing is substantially improved. This is true, however, only when the load is always applied in the same direction. If the load is reversed, the bearing fails instantly. In addition, it should be mentioned that the thicker the oil, the smaller the losses through the pores. On the other hand, a thick oil allows a higher increase in temperature.

Mechanical Conditions

Some mechanical conditions become very important when loads are increased. In order that the pump created by the shaft and the bearing may work properly, the two axes must be perfectly parallel. Any angle between the axes reduces the load capacity of the bearing.

The second important condition is that the bearing must not become oval, which it tends to do under the influence of pressure, for then the maximum bearing load cannot be obtained. This is very important, since the minimum clearance is about 5 μ. Such a distortion reduces the bearing load substantially. A very thick bearing – apart from its oil capacity – will have a smaller bearing load than a rather thin one in a very rigid housing.

Life of the Bearing

In operations in which the oil is kept at a proper temperature (less than 95°C for ordinary oils), the bearing life is determined by both the oil losses and the renewing ability, taking into account the oil inside of the bearing or in a reservoir annex.

Oil losses occur though evaporation, which must not be overlooked especially for electric motors cooled by a powerful ventilator. There may also be accidental losses due to centrifuging by the shaft. When the shaft vibrates transversely, it takes off small quantities of oil which are then centrifuged. To prevent these losses, we have thought of making a groove at each of the bearing so that the oil rings at both ends form inside the bearing (Fig. 5).

As far as the oil-renewing volume which can be extracted form inside the bearing (Fig. 5).

As far as the oil-renewing volume which can be extracted from the bearing is concerned, it can be measured by creating a vacuum inside or outside the bearing while keeping the other side under atmospheric pressure. This experiment shows that the bearing oil content cannot be drawn out completely, and it can be easily understood since the diameters of some pores of the bearing are small enough for the capillary forces to exceed one atmosphere. The corresponding oil volume is lost. In long lasting tests, we took out the bearings after they had failed, and could see that all the oil which could be extracted from the bearing had actually been used. More oil could not be drawn off by creating a vacuum. Consequently, it seems useful to limit the minimum size of the powder particles to about 10 μ.

These slight improvements (the particle selection and the two grooves) cause an ordinary bearing, working under severe conditions, to last 700-1200 hr, instead of 120-140 hr.

Improvements

Apart from the above improvements, the preceding considerations make it obvious that the bearing load would increase if the internal surface was tighter and if air introduction through the pores in the oil film were prevented. That is why we have produced bearings with an internal layer, all the pores of which have a diameter inferior to the value for which oil can be drawn out, thus forming a screen against air introduction.

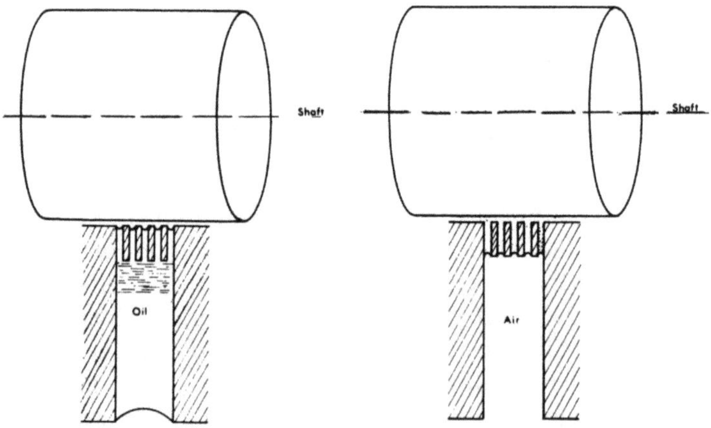

Fig. 6. Schematic representation of a big pore communicating with the fine pores near the shaft before (left) and after (right) the oil film is formed around the shaft.

If R is the pore radius, the capillary pressure for an oil having surface tension of A is given by

$$P = \frac{2A}{R} \tag{3}$$

In the case of mineral oils, the surface tension varies little with temperature and is equal to 32 dynes/cm. The radius R which corresponds to the atmospheric pressure is then 0.64μ (25μin.).

Figure 6 shows how the pores are distributed in a simplified cross section of the bearing. The large pores of the bearing mass open on very small pores of the thin layer near the shaft. In the depression zone the bearing oil will be pumped initially since it will be kept in only by the capillary force of the macropores whose surface tension is low. On the other hand, when the oil is drawn out, the air will not enter because the capillary force will become higher than the atmospheric pressure when the meniscus reaches the thin layer.

The results are indicated by Fig. 7, which shows the variation of the friction coefficient as a function of the load for an ordinary bearing, and for a bearing having a thin layer inside which we have called a "double layer" bearing. Figure 8 gives the maximum value of the maximum PV product when the speed varies.

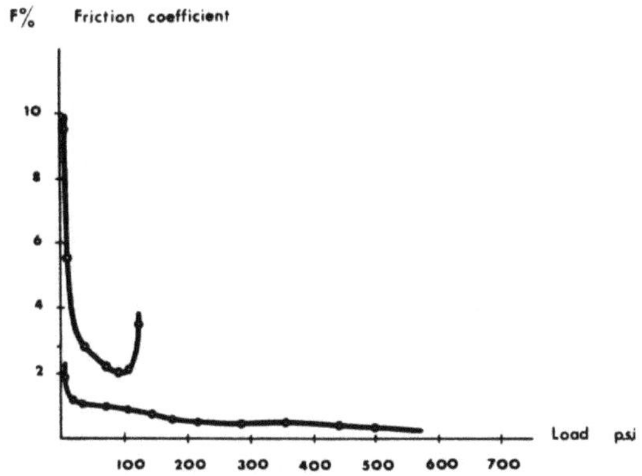

Fig. 7. Friction coefficient as a function of the load for an ordinary bearing (upper curve) and for a "double layer" bearing (below). Note the change of the scale with Fig. 4a (3000 rpm, shaft diameter 1 in., length $^{3}/_{4}$ in.).

Manufacturing

Initially we used an internal layer of carbonyl nickel powder for convenience. Nickel has the advantage of being corrosion-resistant, and its particle size distribution is appropriate. However, when starts are frequent, it has been necessary to add a softer metal. For this reason, a lead and tellurium alloy which

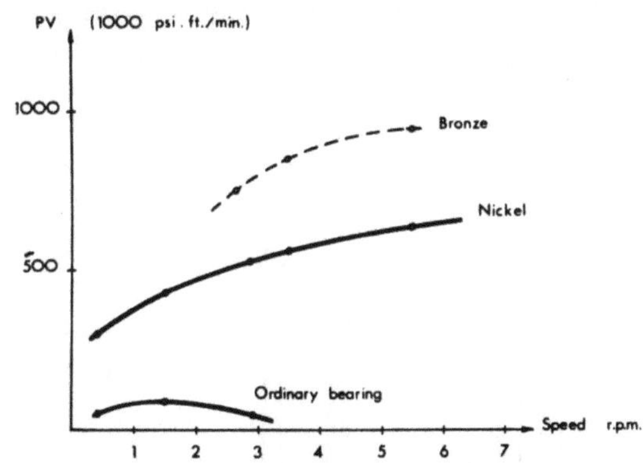

Fig. 8. Product of the speed by the load psi as a function of the number of thousands of runs per minute. Upper curves are related to "double layer" bearings having a fine layer in nickel-tellurium and bismuth (solid line) and bronze (dotted line).

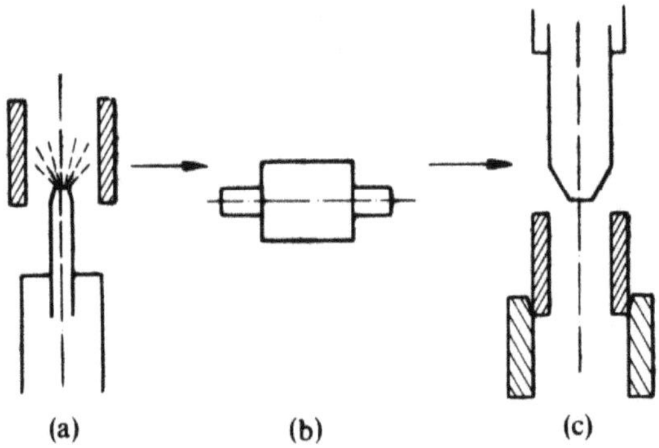

Fig. 9. (a) Spraying of a suspension of a powder in a binder. (b) Sintering. (c) Sizing.

melts only at 917°C has been used. It does not melt at the sintering temperature of the carbonyl nickel, which prevents the formation of macropores. At the present time bearings may be obtained with all the ordinary compositions, such as bronze, iron, brass, aluminum alloys, but bronze with 10% tin has given the best results.

As far as the manufacturing itself is concerned, we have selected two processes which are shown schematically in Figs. 9 and 10. Both use ordinary sintered bearings which are lined inside with a thin layer of a thickness between 0.05 and 0.1 mm. The first process (Fig. 9) consists of spraying the fine powder which is in suspension in an organic binder diluted in a solvent. The

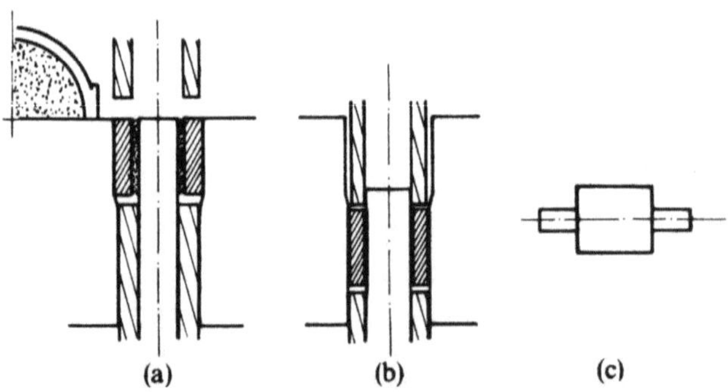

Fig. 10. (a) Filling of the space between the bearing and the core with the fine powder. (b) Compression of the bearing. (c) Sintering.

bearing is then sintered and sized. The second process consists in using the principle of forcing the bearing at the upper part of the die. Then the space between the core and bearing is filled up with the fine powder and, by pressing, the bearing is forced in a narrower part of the die, pulling down the core (mounted on a spring) in such a way that the bearing is pressed around the core. The compression is followed by sintering at low temperature. A further sizing is not necessary. The forcing and filling operations require two steps on a machine equipped with a revolving plate, but they can be made on an ordinary powder metallurgy press.

Conclusions

When considering the results indicated in this paper and in an earlier publication [5], we conclude that under proper operating conditions two causes of rapid failure may be distinguished. The first consists of a progressive elevation of the oil temperature beyond 95-100°C. An improvement may be obtained by reducing the oil viscosity. The second cause of failure is a sudden rupture of the oil film before the maximum temperature is reached, which occurs when the oil is too fluid or when there are too many losses. To lengthen the bearing life, it is necessary to have the largest possible oil reservoir and to prevent losses.

In this paper we did not refer to the different authors who have already cleared up some of the described phenomena; we would like to mention particularly the work of Heuberger [1], Morgan and Cameron [2, 3], and Leloup [4].

References

1. Heuberger, J. H., "Behavior of Lubricants in Porous Bearings," in: W. Leszynski (ed.), Powder Metallurgy, Interscience, New York, 1961, p. 819.
2. Morgan, V. T., and Cameron, A., Conference on Lubrication and Wear, The Institute of Mechanical Engineers, London, 1957, p. 151-157.
3. Cameron, A., Morgan, V. T. and Stainsby, A. E, Proc. Inst. Mech. Engrs. (London), Vol. 176, No. 28, pp. 761-769, 1962.
4. Leloup, L., Études de la lubrication et calcul des paliers, Dunod, Paris, 1964.
5. Eudier, M., Margerand, R., and Youssef, H., Proceedings of the International Powder Metallurgy Conference, Paris, 1964.

Chapter 23

A New Dry-Running Bearing Material

Zdeněk Ministr and Josef Priester

CKD Plant for Semiconductors
Prague, CSSR

Ever since the initial success of the Glacier DU bearings, there has been a mounting interest in (and increasing use of) bearings of this type, which are capable of operating without any lubricant [1,2]. Apart from their excellent performance, they are easy to mass-produce. A steel strip is covered with a porous matrix of powdered tin bronze, and the pores are then filled in with a 4:1 mixture of PTFE with powdered lead. A very thin continuous surface layer of this mixture is provided for running in; once it has been worn away, some 60 to 70% of the bearing surface is formed by the tin bronze.

In a new approach to the problem, we have evolved a new bearing material [3] with a base layer of powdered iron and copper particles, which are interspersed with powdered lead, powdered lead bronze, and graphite. A distinctive feature is the fact that the base consists of only a single layer of particles sintered onto the strip.

Single-Layer Structure

The structures of the conventional DU-type and of the new bearing layers are schematically compared in Fig. 1. In the form-

Fig. 1. Schematic representation of the microstructures of Glacier DU bearings (a) and the new bearing material (b).

er, the several layers of metal particles are relatively closely packed, so that the pores take up no more than 40% at most of the total volume of the layer. In the latter, the particles in the single layer can be much more freely spaced, leaving as much as 65% or even more of the total volume free for the bronze-graphite component.

It is an established fact that in a lubricating layer, the load-bearing capacity is inversely proportional to the thickness. This is another point in favor of the single-layer material, whose thickness is only a fraction of that of the earlier porous types. These could be thinned down by reducing the particle size, but only at the cost of being left with an unacceptably small pore size too. The new material offers a much better ratio of particle size to pore (i.e., void) size.

The pores must be open and form a continuous pattern, because the resistance encountered while the base layer is impregnated with the lubricating component depends on the depth-to-width ratio of the pores, and on their shapes, as well as on the properties of that component. The thin, deep, and tortuous pores of the conventional material will obviously offer a much greater resistance to penetration than the wide and shallow voids between the individual particles of the new material. Their resistance is in fact so great that powdered lead, for instance, cannot be rolled into a porous iron layer at all, for fear of deforming that layer or even the backing strip to which it is sintered. At best, the pores will clog up long before the lead penetrates down to the steel strip. In the new layer, the lead can be driven between the metal particles right down to the steel without any risk of deforming either. This can actually be accomplished with a pressure lower than is needed for perfect compacting of the lead, so that the layer can be made loose enough for bushes to be rounded quite easily from the continuous strip. This could not be done by any other means – not even by sintering a combination of iron and lead powders onto the strip, as the lead would melt and separate the iron particles from the strip. The point is that this prosity enables us to form pliable layers containing nonplastic lamellar substances like graphite or molybdenum disulfide. These layers can be bent down to radii of only a few millimeters, while nonporous layers of the same composition are too brittle to stand bending at all.

Manufacturing Method

Layers of this decription, with a final composition of, e.g., 27% Fe, 26% Cu, 45% Pb, and 2% graphite, are built up successively. First a single layer of FeCu alloy particles, containing some 15% Cu, is sintered onto the steel backing strip so as to leave a porosity of about 65%. The Cu-Pb-graphite component is then introduced (e.g., rolled) into the voids loosely enough to allow for its further compacting when the bearings are rounded and their internal surfaces are calibrated to final size.

At present, bushes of 20 to 30 mm diameter, with a wall thickness of 1.5 mm (including a lubricating layer 0.2 mm thick), are manufactured by the following procedure. An acid-pickled 1.3-mm strip of steel is covered with a single layer of a 10:1 mixture of FeCu particles and electrolytic copper powder. The FeCu alloy is prepared by annealing 85 parts of atomized iron with an 0.15- to 0.25-mm grain size together with 15 parts of electrolytic copper powder. The layer is sintered for 7 or 8 minutes at 1120 to 1150°C in a reducing atmosphere. It is then lightly rolled down to a thickness of 0.25 mm and checked for the volume of metal in it (which should take up about 35% of the total layer volume). A mixture of powdered lead bronze, powdered lead, and graphite is next rolled into the voids, and residual stresses are relieved by a short spell of annealing at 260 to 280°C in a reducing atmosphere. The strip is then cut into the requisite lengths, and these are rounded into bushes (which reduces the porosity to 3-5%) and calibrated to size (which brings the porosity down to 2-3%).

Ratio of Thickness

The pliability of the finished strips can be approximated as

$$\Delta/d = (2v + s)/d$$

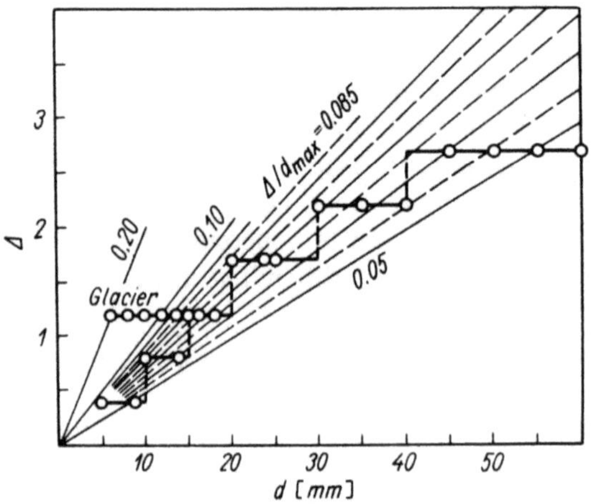

Fig. 2. The dependence of Δ upon *d*, and the range of wall thicknesses, in the new bearings.

where Δ is the relative shortening of the sliding surface after rounding, *v* the thickness of that layer, *s* the thickness of the steel strip, and *d* the bore of the bush.

The dependence of Δ on *d* is plotted in Fig. 2. It should be noted that while the Glacier DU strips can be bent to radii where Δ:*d* is 0.10 to 0.20 (the latter value is attained with a 6-mm mandrel), the new graphite-containing layers will bend into 20-mm bores only if the Δ:*d* ratio is below 0.08. At Δ:*d* equal to or higher than 0.085, layers with 2% of graphite and a 2-3% porosity were apt to crack, while similar layers with a porosity of 3-5% remained pliable. This led to the realization that mandrels of less than 20 mm diameter called for special ratios of layer thickness to strip thickness.

There is no purpose in building up layers whose thickness is a multiple of the permissible bearing wear, which in turn is only rarely more than five times the initial clearance. In the new type of bearings, it has been found advisable to keep this initial clearance to 0.2% *d*, so that the layer thickness should be not more than 1% *d* — e.g., 0.2 mm for *d* = 20 mm. Thicker layers would add to the weight of the bearing without in any way improving its strength or performance. As there are limits to the pliability of the layer, the steel base strip should likewise be as thin as possible – only just thick enough to afford a secure fit of the bearing. Since production economics call for the least possible number of different wall thicknesses ($v + s$), it has been found convenient to confine these to a few steps as shown in Fig. 2. For bores above 15 mm, the new bearings can have the same wall thickness as the equivalent Glacier DU type, but for smaller diameters they must always be thinner.

Wear Tests

The p versus v and f versus v curves (where p is the specific pressure, f the coefficient of friction, and v the circumferential velocity) were examined in tests where the new bearings and Glacier DU bearings were run side by side, unlubricated, under identical conditions. When the static load was gradually increased until the bearing temperature was stabilized at 100°C, closely similar p versus v (load-capacity) curves were obtained from both types of bearings. (The differences between them lay within the range of measurement errors.) Greater differences were found in the f versus v curves. In the new bearings f appeared to be around 0.1, and in the DU bearings about 0.2. Apart from the graphite component, however, the low friction of the new bearings may have been influenced by traces of lubricant left in the pores by the coining operation.

To investigate the wear resistance, the bearings of both types were fitted to standard starter motors normally used for 1000-cc car engines, and subjected to simulated starts (i.e., to shock loads) under conditions more severe than any likely to be encountered in practice. There are two bearings to each starter. The one at the pinion end has to sustain the impacts and is more exposed to the ingress of dirt, but is better cooled than that at the commutator end. No attempt was made to simulate the effects of dirt from the road, but the pinion-end bearings were left exposed to metal particles abraded off the pinion and the toothed ring that meshes with it. The starters were tested six at a time, three with each type of bearing. The pinion-end DU bearings became unserviceable after an average of 7000 engine starts, while the new bearings were still capable of further operation after 25,000. By that time their clearance had increased by 0.1 mm, which means that half the layer had been worn away. On the whole, the DU bearings at the pinion ends wore 5.4 times as fast, and those at the commutator ends 3.5 times as fast, as bearings of the new type.

These bearings thus offer roughly the same load-bearing capacity as the DU type, but afford a markedly lower coefficient of friction and substantially higher wear resistance.

References

1. British Patent No. 657,080.
2. British Patent No. 657,085.
3. Czechoslovak patent application PV 918-64. Also patented in Austria (249, 455), France (1, 428,120), and patents applied for in the GDR, GFR, Great Britian, Italy, USA, and USSR.

Chapter 24

Cost Structure for Self-Lubricating Bearings

O. Wiberg

Svenska Metallverken
Västerås, Sweden

Introduction

The author's firm, Svenska Metallverken, chiefly manufactures semifinished products from nonferrous metals. The head office and main works dealing with copper and copper alloys are at Västerås, Sweden. The total personnel is about 7000, of whom 1900 are employed at the Västerås works. The powder-metallurgy department there is organized as a separate unit, covering manufacture, development, and sales, under a production manager, and employing 35 workmen and 17 office staff.

Production Costs

The main powder-metallurgy product is self-lubricating bearings which accounts for more than 80 vol.% of the output. Of the 200 tons/year produced, 85% consists of bronze bearings and the remainder of iron. The average weight of the parts is 23 g. All tools are bought from outside.

Overheads

The method of tabulating costs varies from country to country and from firm to firm. An attempt has been made here to follow the "Account Classification Guide" prepared by the Powder Metallurgy Parts Association (USA), and the following data have been compiled on this basis.

	Percentage of total
Raw material	41.2
Direct labor	7.5
Factory running expenses	31.0
Administration	10.3
Sales	10.0
	100.0

These figures relate to the year 1964, and the exceptionally high figure for raw material is explained by the heavy increase in price of copper and tin during that period. The cost of direct labor is low, despite an average hourly wage of $1.34, excluding all social benefits, which amount to 20%. The high "sales" cost is typical of the Scandinavian market, which is largely made up of many small customers, scattered over a wide area.

Costing of a Specific Bearing

It is almost impossible to arrive at the actual cost of production of a specific part, so that it is necessary to devise some formula upon which prices can be based. Each firm has its own system, often reduced to a form as simple as

$$X + Y \times Z = \text{total cost}$$

where X is the raw material, Y is the direct labor cost, and Z is a factor embracing all other items. In the case of Svenska Metallverken, the factor Z amounts to ~8.

This method of calculation gives very misleading results. For a given part, a low labor cost is generally the outcome of more expensive tooling, more expensive machinery, and longer engineering time. It is thus almost impossible to determine Z in a specific case.

Table 1 presents a breakdown of the costs of manufacture of 100,000 bronze bearings in a few typical cases where the orders received vary somewhat:

Table 1. Total Cost (U. S. Dollars) of Producing 100,000 Bronze Bearings, Depending upon Type of Order

Case	Tools	Tool setup*	Raw material†	All other costs	Total cost	Cost per 100	Increase over A, %
A	250 (2.7%)	20 (0.2%)	3780‡ (41.2%)	5150 (56.9%)	9200 (100.0%)	9.20	0
B	250 (2.6%)	200 (2.1%)	3830§ (40.8%)	5150 (54.4%)	9430 (100.0%)	9.43	2.5
C	2500 (21.4%)	200 (1.7%)	3830§ (32.6%)	5150 (44.3%)	11,680 (100.0%)	11.68	27.0
D	2500 (21.0%)	400 (3.4%)	3830§ (32.2%)	5150 (43.4%)	11,880 (100.0%)	11.88	29.2

* The costs of tool setup include maintenance of tools and cost of changing the tools, plus idle time in the presses.

† Based on Cu 90/Sn 10 at 49 cents/lb for copper and 153 cents/lb for tin.

‡ Yield 99%.

§ Yield 98%.

Case A: 100,000 bronze bearings, weight 23 g each, in one shipment. New tools worth $250 required.

Case B: Same as A, but order for 10 shipments.

Case C: 10 different but similar bearings (23 g); one shipment comprising 10,000 bearings to each specification. 10 new tools worth $250 each have to be made.

Case D: Same as C, but to be despatched in two shipments of 5000 bearings to each specification.

Conclusions

Knowing that most small and medium-sized plants in this branch of industry are running at a net profit before tax of 5-10%, it is easy to see that a weak sales policy will severely damage the business.

If a customer starts to split his large orders into small shipments (Case A vs. Case B) this will mean a substantial reduction in profits. A really dangerous step is to gamble on the price of tools anticipating future large orders (Case A vs. Case C). If one out of three customers fails to repeat his order, the overall position will show a loss.

The present rapid fluctuations on the metal market render it impossible to remain in business without making provision for changes in raw material prices. A more normal figure for raw materials would be 30%. Running orders for bearings at fixed prices was certainly not profitable in 1964.

Case D represents the type of order most commonly received. A price difference over Case A of 29·2% is necessary to compensate for the higher costs. To convince the customer of this fact remains a major problem.

Appendix

METAL POWDER INDUSTRIES FEDERATION

P/M Materials Standards and Specifications*

MPIF Standard 35
Adopted 1961, Revised 1965, 1969

Scope

MPIF Standard 35 was issued to provide the design and materials engineer with the information necessary for specifying those P/M materials which have been developed and accepted by the P/M parts manufacturing industry as standard and representative of their capabilities and commercial practices. It is reasonable to expect that any properly equipped and experienced P/M parts manufacturer could supply P/M materials that will meet this standard.

MPIF Standards are issued and adopted in the public interest. They are designed to eliminate misunderstandings between the manufacturer and the purchaser and to assist the purchaser in

*This Standard, prepared by the Metal Powder Industries Federation, is subject to periodic revision. Suggestions for revision should be addressed to the Metal Powder Industries Federation, 201 East 42nd Street, New York, N.Y. 10017. Users of standards are cautioned to secure the latest editions. Additional information must be approved by the Standards Board of the Metal Powder Industries Federation before it can be considered part of the Standard. Extra copies of MPIF Standards may be obtained from the Federation at above address. A list of the other standards and prices will be sent on request.

selecting and obtaining the proper material for his particular product. Existence of an MPIF Standard does not in any respect preclude any member or nonmember from manufacturing or selling products not included in this Standard.

A Standard of the Metal Powder Industries Federation defines a product, process, or procedure with reference to one or more of the following: nomenclature, composition, construction, physical properties, mechanical properties, dimension, tolerances, safety, operating characteristics, performance, quality, rating, testing, and the service for which designed.

The data provided herein represent typical mechanical properties achieved under conventional manufacturing procedures. Higher mechanical properties and other improvements in performance characteristics can be obtained by experienced custom P/M parts manufacturers through the use of more complex processing techniques or different starting materials.

By referring to the property and performance values specified in each of the P/M Materials Standards, it is possible for the user to designate the exact material he desires for a specific application. It should also be apparent that the range of properties available through powder metallurgy is constantly increasing and that there usually are several alternative approaches toward achieving identical properties and performance. The data given are realistic and can serve as an effective guideline in the proper application of the products of powder metallurgy.

Explanatory Notes

The following describe and explain the details of the P/M Materials Standards. An understanding of what is intended will aid the Standards user in the proper interpretation of the data furnished.

Material. A word description of the P/M material being specified.

Designation. A code designation of the P/M material being specified which is based on the P/M coding system adopted by the industry. The coding system offers a convenient means for designating the type of material, composition, and density of a P/M part. It should be noted that those code designations shown in Standard 35 and revisions thereof apply only to P/M materials standardized by the Metal Powder Industries Federation. (Note: Materials not yet adopted as industry standards should not be classified under the MPIF coding system.)

In the coding system the prefix letters denote the general type of material; for example, the letters "BT" = bronze.

Prefix Letter Code

BT	Bronze
BTP	Bronze (leaded)
BZ	Brass
BZN	Nickel silver
BZNP	Nickel silver (leaded)
BZP	Brass (leaded
F	Iron (or steel)
FC	Copper steel (or iron-copper)
FN	Nickel steel (or iron-nickel)
FX	Infiltrated steel (or infiltrated iron)
SS	Stainless steel

The four digits following the prefix refer to the composition of the material.

In nonferrous materials, the last two numbers in the four-digit series designate the precentage of the major alloying constituent. If there are no minor constituents, two zeros are indicated for the first two digits. If a minor constituent is present, the first two numbers designate the percentage of the minor constituent. Additional minor constituents are indicated in the prefix letters only.

In ferrous materials, those major alloying elements (except carbon) which are shown as ranges are included in the prefix letter code. Other elements are excluded from the code but are represented in "Chemical Composition" (Line 1). The first two digits of the designation indicate the percentage of the major alloying constituent present.

Carbon content in ferrous materials is designated by the last two numbers in the four-digit series. The carbon content up to and including 0.3% will be considered as zero, higher contents will be indicated in ranges and coded as follows:

Carbon ranges	Code designation
0.0%-0.3%	00
0.3%-0.6%	05
0.6%-1.0%	08

The percentage of carbon that is metallurgically combined is to be indicated in the coding system. Free carbon such as graphite in certain bearings is not to be included in the coding but may be referred to elsewhere in the Standard.

In the case of P/M Stainless Steels the four-digit series shall be replaced with the appropriate designation adopted by the American Iron and Steel Institute, but the MPIF letter codes will still be used as part of the designation in this standard.

The suffix letter denotes the density range of the material. The range used is that in which the mean density lies.

Suffix Letter Codes for Density Range Classification

N . . . . Less than 6.0 g/cc
P . . . . 6.0 to less than 6.4 g/cc
R . . . . 6.4 to less than 6.8 g/cc
S . . . . 6.8 to less than 7.2 g/cc
T . . . . 7.2 to less than 7.6 g/cc
U . . . . 7.6 to less than 8.0 g/cc
W . . . . 8.0 to less than 8.4 g/cc

Following are line-by-line explanatory details for each of the items included in the P/M Materials Standard.

Chemical Composition (Line 1)

Elements are listed as minimum and maximum percentage. "Other Elements" includes the total other elements by difference and is reported as a maximum percentage. The compositions are those of the basic material before any oil impregnation, infiltration, heat treatment, etc. has taken place.

Typical P/M Material Code Designations

Material	Composition, %	Code for material and composition
	Nonferrous	
P/M bronze	Cu-88, Sn-10	BT-0010
P/M bronze (leaded)	Cu-87, Sn-10, Pb-3	BTP-0310
P/M nickel silver	Cu-64, Zn-18, Ni-18	BZN-1818
P/M nickel silver (leaded)	Cu-64, Ni-18, Zn-16	BZNP-1618
P/M brass (leaded)	Cu-79, Zn-18, Pb-2	BZP-0218
	Ferrous	
P/M iron	Fe-99, C-0.2	F-0000
P/M steel	Fe-98, C-0.8	F-0008
P/M copper steel	Fe-96, Cu-2, C-0.8	FC-0208
P/M iron copper	Fe-89, Cu-10, C-0.2	FC-1000
P/M nickel steel	Fe-96, Ni-2, C-0.2	FN-0200
P/M infiltrated iron	Fe-78, Cu-20, C-0.2	FX-2000
P/M infiltrated steel	Fe-77, Cu-20, C-0.8	FX-2008
P/M austenitic stainless steel	AISI 316	SS-316
P/M martensitic stainless steel	AISI 410	SS-410

Condition (Line 2)

The properties of a P/M material are normally a function of its condition, i.e., as-sintered (column 2) or heat treated (column 3), or subjected to any other special processing (column 4). As-sintered refers to the normal condition of P/M parts after final compacting and sintering. Data shown under columns 3 and 4 will be identified as to condition.

Density (Line 3)

Density is expressed in grams per cubic centimeter. Dry density is the weight per unit volume of an unimpregnated P/M part. Wet density is the weight per unit volume of a P/M part impregnated with oil or other nonmetallic materials. Normally, density of bearings is reported on a fully impregnated basis, and density of mechanical components on a dry, unimpregnated basis.

Density may be calculated by any of several means, provided the method is mutually agreed upon by manufacturer and purchaser.

A method commonly used is as follows:

Mechanical components:

$$D = \frac{A}{B-C}$$

Oil-impregnated bearings

$$D = \frac{B}{B-C}$$

where

D = density, in grams per cubic centimeter,
A = weight of the unimpregnated sample in air, in grams,
B = weight of the oil-impregnated sample in air, in grams, and
C = weight of the oil-impregnated sample immersed in water, in grams.

Note 1. Weights A, B, and C shall be determined to within 0.1%.
Note 2. When bearings are purchased impregnated, they should be weighed as received for determining weights of B and C.
Note 3. The effect of the surface tension of water in weighing the test sample may be minimized by the addition of a wetting agent to the water.

When a sample is to be impregnated for the purpose of determining weights B and C, either of the following two procedures may be used; however, Procedure No. 1 is the preferred method and shall be used unless the manufacturer and purchaser agree to use Procedure No. 2. Procedure No. 2 might result in impregnation as much as 25% less than obtained by Procedure No. 1.

(1) The pressure over the sample after immersion in oil (viscosity of approximately 200 sec Saybolt Universal at 100°F and/or as agreed between manufacturer and purchaser) at room temperature shall be reduced to not more than 2 in.Hg pressure for 30 min by a suitable evacuating method, after which the pressure shall be increased to atmospheric pressure for 10 min, with the

sample immersed in the oil at room temperature during the entire period.

(2) The sample shall be immersed for at least 4 hr in oil, held at a temperature of 180 ± 10°F, and then cooled to room temperature by immersion in oil at room temperature.

Samples that are to be freed of lubricant shall be extracted in Soxhlet apparatus using a suitable solvent, such as petroleum ether. After extraction residual solvent shall be removed by heating sample at 250°F for 1 hr. Alternate extraction and drying shall be continued until the dry weight is constant to 0.1%. For further details refer to ASTM Standard B 328.

Allowable Density Variation (Line 4)

The allowable density variation is the maximum allowable variation in density for a given material, usually expressed in grams per cubic centimeter. In mechanical components, if the density varies from one section to another more than the stated allowable variation, the manufacturer and the purchaser shall agree upon a critical section of the component where the stresses are highest. The density of this critical section rather than the average density shall fall within the limits described in the P/M Materials Standard.

Interconnected Porosity (Line 5)

Interconnected Porosity in a P/M material is that network of pores in and extending to the surface of the sintered compact. It is controllable and a function of the raw material used and the manufacturing techniques applied. Interconnected Porosity for mechanical components is not normally part of the specification, unless requested by the purchaser. The pore volume or oil content of self-lubricating P/M bearings, in terms of Interconnected Porosity, shall be calculated as follows:

$$P = \frac{B-A}{(B-C) \times S} \times 100$$

where

P = oil content or pore volume, in percent,
A = weight of unimpregnated sample in air, in grams,
B = weight of oil-impregnated sample in air, in grams,

C = weight of oil-impregnated sample immersed in water, in grams, and

S = specific gravity of impregnant at the temperature of test.

Note 1. Weights A, B, and C shall be determined to the nearest 0.1%.

Note 2. When bearings are purchased impregnated they should be weighed as received for determining values of B and C.

Note 3. Samples that are to be freed of lubricant shall be extracted in Soxhlet apparatus using a suitable solvent, such as petroleum ether. After extraction residual solvent shall be removed by heating sample at 250°F for 1 hr. Alternate extraction and drying shall be continued until the dry weight is constant to 0.1%. Surface porosity of oil-impregnated bearings may be determined by heating the sample as received to a temperature not exceeding 300°F for a period of not more than 5 min, during which time oil shall exude uniformly from the surface.

For further details refer to ASTM Standard B 328.

Properties (Line 6-15 and 17-26)

Lines 6 through 15 and lines 17 through 26 provide property data where available. It should be noted that in most instances the data merely indicate to the purchaser typical values that may be expected from special sub-size test specimens conforming to the chemical and density requirements indicated. All strength and elongation values are obtained from MPIF standard test bars and not from samples cut from commercial components. Data are not available for all properties listed on the Materials Standards but may be added in subsequent revisions as it becomes available or necessary.

The manufacturer and the purchaser may agree on qualification tests for functional performance. These should be made on sample components and the tests should be determined after consideration of the function the component is to perform.

Ultimate Tensile Strength (Line 7)

Ultimate Tensile Strength is expressed in pounds per square inch and is the ability of a P/M part to resist fracture when a pulling force is applied in the direction parallel to its longitudinal axis. The Ultimate Tensile Strength is equal to the maximum load divided by the original cross-sectional area. Standard tension test specimens have been established by the Metal Powder Industries Federation (MPIF Standard 10). For further details refer to ASTM Standard E8.

Tensile Yield Strength (Line 8)

Tensile Yield Strength is expressed in pounds per square inch. It is the stress at which a material exhibits a specified deviation from proportionality of stress and strain in tension. Reading shall be taken at an offset of 0.2%. Standard tension test specimens have been established by the Metal Powder Industries Federation (MPIF Standard 10). For further details refer to ASTM Standard E8.

Compressive Yield Strength (Line 9)

Compressive Yield Strength is expressed in pounds per square inch. It is the stress at which a material exhibits a specified deviation from proportionality of stress and strain in compression. Readings shall be taken at an offset of 0.1%. Standard compression specimens have not been developed by the MPIF. For further details refer to ASTM Standard E9.

Elongation (Line 10)

Elongation is expressed as a percentage of the original gauge length and based on measuring the increase in the gauge length after fracture of the specimen in tension, providing the fracture takes place within the gauge length. For further details see ASTM Standard E8.

Apparent Hardness (Line 11)

Apparent Hardness is the value obtained by testing a sintered P/M material with standard indentation hardness equipment. Since the reading is a composite of pores and solid material, it is usually lower than that of solid material of the same composition and condition. It should not be confused with matrix hardness which is the true hardness of a P/M material, as determined by microhardness equipment. Apparent hardness values are often agreed upon between the manufacturer and purchaser. The effect of the density distribution within the component must be considered when establishing hardness limits.

"K" Strength Constant (Line 12)

The "K" Strength Constant is used in the determination of radial crushing strength of bearings. In P/M bearings, the radial crushing strength shall not be less than the value calculated as follows:

$$P = \frac{KLT^2}{D-T}$$

where

P = radial crushing strength, in pounds,
D = outside diameter of bearing, in inches,
T = wall thickness of bearing, in inches,
L = bearing length, in inches, and
K = strength constant as set forth in P/M Materials Standard.

Radial crushing strength shall be determined by compressing the test bearing between two flat surfaces, the direction of the load being normal to the longitudinal axis of the bearing. The point at which the load drops due to the first crack shall be considered the crushing strength. This test shall be applied to plain cylindrical bearings. Flanged bearings shall be tested by cutting off the flange and compressing the two sections separately. Each section shall meet the minimum strength requirements prescribed above.

Impact Strength (Line 13)

Impact Strength is expressed in foot-pounds, and is a measurement of the amount of energy needed to fracture a specimen with a single blow. Two impact strength tests are used in powder metallurgy, the Charpy test and the Izod test.

In each case the specimen shall be tested in accordance with customary procedures as set forth in the ASTM Standard E 23. The impact strength data shown in the P/M Materials Standard were derived by striking unnotched specimens perpendicular to the direction of the pressing.

Fatigue Strength (Line 14)

Fatigue Strength is expressed in pounds per square inch, and represents the maximum stress that can be sustained for a specified number of cycles without failure. The method of test is described on line 28 where applicable.

Special Properties (Line 15)

Because of the diverse nature of application for P/M materials, and the wide compositional ranges available, P/M materials can often be provided with special properties such as magnetic or electrical properties. If pertinent to the use of the material specified, such special properties may be described in Line 15.

Typical Applications (Line 16)

Describes the nature of the recommended applications for the specific material.

Young's Modulus (Line 18)

Young's Modulus is the ratio of normal stress to corresponding strain for tensile or compressive stresses below the proportional limit of the material, expressed in pounds per square inch. For further details see ASTM Standard E 111.

Poisson's Ratio (Line 19)

Poisson's Ratio is the absolute value of the ratio of transverse strain to the corresponding axial strain resulting from uniformly distributed axial stress below the proportional limit of the material.

See ASTM Standard E 132 "Determination of Poisson's Ratio at Room Temperature" for more details.

Additional Properties (Line 20 through 26)

This space shall be used for additional property data when developed and adopted by the industry.

Comparable Specifications (Line 27)

For cross-reference purposes, comparable standards issued by other standards-promulgating institutions are listed here. Note that the absence of a listing does not necessarily mean that a comparable standard does not exist. To procure copies of reference standards contact:

American Society for Testing and Materials
1916 Race Street
Philadelphia, Pennsylvania 19103

Society of Automotive Engineers
2 Pennsylvania Plaza
New York, New York 10001

For military standards contact the Superintendent of Documents, U. S. Government Printing Office, Washington, D. C. 20402.

Comments (Line 28)

Space is provided here for any comments or special explanatory notes that apply specifically to the material described in the Standard.

GENERAL INFORMATION AND RECOMMENDED PRACTICES FOR P/M PARTS

Lot. Unless otherwise specified, a lot shall consist of pieces of the same form and dimensions made from powders of the same composition, molded and sintered under the same conditions, and submitted for sampling and inspection at one time.

Samples for Chemical Analysis. At least one sample for chemical analysis shall be taken from each lot. A sample shall consist of chips obtained by milling, drilling, or crushing from at least two pieces with clean, dry tools without lubrication. In order to obtain oil-free chips, the parts selected for test shall be extracted as described under Density (line 3).

Workmanship. Mechanical components and bearings shall be uniform in composition and free from defects which would affect their serviceability.

Variations in Dimensions. Variations in dimensions shall be withing the limits specified on the drawings which describe the mechanical components and/or bearings, and which accompany the order; or variations shall be within the limits specified in the order.

Inspection. Unless otherwise specified, inspection of mechanical components and/or bearings supplied on contract shall be made by the purchaser.

Certification. A certification, based on manufacturer's quality control, that the material conforms to the requirements

Table I. Permissible Loads

Shaft velocity, fpm	Permissible loads, psi				
	BT-0010-N	BT-0010-R	BT-0010-S	F-0000-N	FC-1000-N
Static	5,500 max	8,500 max	11,250 max	7,500 max	15,000 max
Slow and intermittent	3,200	4,000	4,500	3,600	8,000
25	2,000	2,000	1,800	1,800	3,000
50 to 100	550	500	450	400	700
100 to 150	365	325	300	235	400
150 to 200	280	250	225	175	300
Over 200	$P = \frac{55,000^*}{V}$	$P = \frac{50,000}{V}$	$P = \frac{45,000}{V}$	$P = \frac{35,000}{V}$	$P = \frac{40,000}{V}$

* For shaft velocities of 500 to 1000 fpm, P = 60,000/V (for BT-0010-N), where P is the safe load in pounds per square inch of projected bearing area (length times inside diameter of bearing), and V is the shaft velocity in ft. per min.

of these specifications, shall be the basis of shipment of the material. A certificate covering conformance of the material to these specifications shall be furnished by the manufacturer upon request of the purchaser.

Rejection. Rejection based on tests made in accordance with these specifications shall be reported to the manufacturer within the period agreed upon between the manufacturer and the purchaser.

RECOMMENDED PRACTICES FOR P/M BEARINGS

Permissible Loads. In calculating permissible loads, the operating conditions, housing conditions, and construction should be considered. Permissible loads for various operating conditions are shown in Table I. These are intended only as a general guide. It is recommended that all new applications be tested prior to establishing purchasing specifications.

Certain conditions will increase the permissible loads; such as additional lubrication, pressure lubrication, cooling, hardening or chrome-plating the shaft, and loads of short duration.

Certain conditions tend to reduce the load-carrying capacity of bearings regardless of type or make: continual start-stop operation, oscillatory or reciprocating motion; high speeds; shock loading; extremely high or low temperatures; excessively close or

Table II. Recommended Press Fits

Outside diameter, in.	Press fit, in.	
	min	max
Up to 0.760	0.001	0.003
0.761 to 1.510	0.0015	0.004
1.511 to 2.510	0.002	0.005
2.511 to 3.010	0.002	0.006
Over 3.010	0.002	0.007

loose bearing clearances; shaft run out, deflection or misalignment of shaft; dust, grit, moisture, corrosive fumes, or poor shaft finish; oil breakdown.

To reduce the unit loading "P" on a bearing, the length must be increased (see Table IV for limits) because any increase in diameter is offset by an increase in the shaft velocity.

The permissible loading on thrust bearings is approximately 20% of that for plain bearings. An adequate oil film, as provided by immersion in oil or pressure lubrication, will increase the permissible loading.

Press Fits (Interference Fits). Plain cylindrical journal bearings are commonly installed by press fitting the bearing into a housing with an insertion arbor. For housings rigid enough to withstand the press fit without appreciable distortion and for bearings with wall thickness approximately one-eighth of the bearing outside diameter, the press fits shown in Table II are recommended.

Running Clearance. Proper running clearance for bearings depends to a great extent on the particular application. Only min-

Table III. Running Clearances

Shaft size, in.	Clearance, min, in.	
	Bronze base	Iron base
Up to 0.760	0.0005	0.001
0.761 to 1.510	0.001	0.0015
1.511 to 2.510	0.0015	0.002
Over 2.510	0.002	0.0025

Table IV. Dimensional Tolerances for Plain Cylindrical Bearings

Inside diameter and outside diameter, in.	Total diameter tolerance, in.*			
	Bronze base		Iron base	
	Inside diameter	Outside diameter	Inside diameter	Outside diameter
Up to 0.760	0.001	0.001	0.001	0.001
0.761 to 1.010	0.001	0.001	0.0015	0.0015
1.011 to 1.510	0.001	0.001	0.0015	0.0015
1.511 to 2.010	0.0015	0.0015	0.002	0.002
2.011 to 2.510	0.0015	0.0015	0.002	0.002
2.511 to 3.010	0.002	0.002	0.003	0.003
3.011 to 4.010	0.003	0.003	0.004	0.004
4.011 to 5.010	0.004	0.004	0.005	0.005
5.011 to 6.010	0.005	0.005	0.006	0.006

Length, in.	Total length tolerance, in.†	
	Bronze base	Iron base
Under 1.495	0.010	0.010
1.496 to 1.990	0.015	0.015
1.991 to 2.540	0.015	0.020
2.541 to 2.990	0.015	0.020
2.991 to 3.985	0.020	0.030
3.986 to 4.985	0.020	0.030

Outside diameter, in.	Wall thickness, max, in.	Concentricity tolerance, in. ‡	
		Bronze base	Iron base
Up to 1.010	Up to 0.255	0.003	0.003
1.011 to 1.510	Up to 0.355	0.003	0.003
1.511 to 2.010	Up to 0.505	0.004	0.004
2.011 to 3.010	Up to 0.760	0.005	0.005
3.011 to 4.010	Up to 1.010	0.005	0.005
4.011 to 5.010	Up to 1.510	0.006	0.006
5.011 to 6.010	Up to 2.010	0.007	0.007

* Total tolerance on the inside diameter and outside diameter is a minus tolerance only.

†Total tolerance is split into plus and minus.

‡Total indicator reading.

Table V. Dimensional Tolerances for Flange Bearings and Thrust Washers

Flange bearings: flange diameter, in.	Flange diameter tolerance, in.			
	Bronze base		Iron base	
	Class A	Class B	Class A	Class B
0 to $1^{1}/_{2}$	±0.0025	±0.005	±0.0025	±0.005
$1^{1}/_{2}$ to 3	±0.005	±0.010	±0.005	±0.010
3 to 6	±0.010	±0.025	±0.010	±0.025

Note: Normally the outside diameter of the flange is not too critical; therefore, should not be held too close – unless required. Class A tolerances may require additional operations.

Flange bearings: flange diameter, in.	Flange thickness tolerance, in.			
	Bronze base		Iron base	
	Class A	Class B	Class A	Class B
0 to $1^{1}/_{2}$	±0.0025	±0.005	±0.0025	±0.005
$1^{1}/_{2}$ to 3	±0.007	±0.010	±0.007	±0.010
3 to 6	±0.010	±0.015	±0.010	±0.015

Note: Normally the thickness of the flange is not too critical; therefore, should not be held too close– unless required. Class A tolerances may require additional operations.

Thrust washers: thickness, in.	Thickness tolerance, in. (for all diameters)			
	Bronze base		Iron base	
	Class A	Class B	Class A	Class B
0 to $^{1}/_{4}$	±0.0025	±0.005	±0.0025	±0.005

Flange bearings and thrust washers: diameter, in.	Parallelism on faces, max, in.			
	Bronze base		Iron base	
	Class A	Class B	Class A	Class B
0 to $1^{1}/_{2}$	0.002	0.003	0.003	0.005
$1^{1}/_{2}$ to 3	0.003	0.004	0.005	0.007
3 to 6	0.004	0.005	0.007	0.010

Note: Closer tolerances can be held, but may require additional operations.

Note: For flange bearings, the body tolerances – inside diameter, outside diameter, length, and concentricity – are the same as for plain cylindrical bearings, Table V.

Table VI. Inch – Millimeter Equivalents

in.	0	1	2	3	4	5	6	7	8	9
	mm									
0. . .	. . .	25.4	50.8	76.2	101.6	127.0	152.4	177.8	203.2	228.6
10. . .	254.0	279.4	304.8	330.2	355.6	381.0	406.4	431.8	457.2	482.6
20. . .	508.0	533.4	558.8	584.2	609.6	635.0	660.4	685.8	711.2	736.6
30. . .	762.0	787.4	812.8	838.2	863.6	889.0	914.4	939.8	965.2	990.6
40. . .	1016.0	1041.4	1066.8	1092.2	1117.6	1143.0	1168.4	1193.8	1219.2	1244.6
50. . .	1270.0	1295.4	1320.8	1346.2	1371.6	1397.0	1422.4	1447.8	1473.2	1498.6
60. . .	1524.0	1549.4	1574.8	1600.2	1625.6	1651.0	1676.4	1701.8	1727.2	1752.6
70. . .	1778.0	1803.4	1828.8	1854.2	1879.6	1905.0	1930.4	1955.8	1981.2	2006.6
80. . .	2032.0	2057.4	2082.8	2108.2	2133.6	2159.0	2184.4	2209.8	2235.2	2260.6
90. . .	2286.0	2311.4	2336.8	2362.2	2387.6	2413.0	2438.4	2463.8	2489.2	2514.6
100. . .	2540.0	. . .	. . .	. . .	. . .	. . .	. . .	. . .	. . .	. . .

Note: All values in this table are exact, based on the conversion factor 1 in. = 25.4 mm (exactly). By manipulation of the decimal point any decimal value or multiple of an inch may be converted to its exact equivalent in millimeters.

Table VII. Pressure Stress Equivalents (Thousand pounds-force per square inch to kilograms-force per square millimeter)

Conversion factors: 1 in. = 25.4 mm (exactly)
1 lbf = 0.453 592 37 kgf (exactly)

ksi	0	1	2	3	4	5	6	7	8	9
	kgf/mm^2									
0. .	. . .	0.703 1	1.4061	2.1092	2.8123	3.5153	4.2184	4.9215	5.6246	6.3276
10. .	7.0307	7.733 8	8.4368	9.1399	9.8430	10.5460	11.2491	11.9522	12.6553	13.3583
20. .	14.0614	14.764 5	15.4675	16.1706	16.8737	17.5767	18.2798	18.9829	19.6859	20.3890
30. .	21.0921	21.795 2	22.4982	23.2013	23.9044	24.6074	25.3105	26.0136	26.7166	27.4197
40. .	28.1228	28.825 9	29.5289	30.2320	30.9351	31.6381	32.3412	33.0443	33.7473	34.4504
50. .	35.1535	35.856 5	36.5596	37.2627	37.9658	38.6688	39.3719	40.0750	40.7780	41.4811
60. .	42.1842	42.887 2	43.5903	44.2934	44.9965	45.6995	46.4026	47.1057	47.8087	48.5118
70. .	49.2149	49.917 7	50.6210	51.3241	52.0271	52.7302	53.4333	54.1364	54.8394	55.5425
80. .	56.2456	56.948 6	57.6517	58.3548	59.0578	59.7609	60.4640	61.1671	61.8701	62.5732
90. .	63.2763	63.979 3	64.6824	65.3855	66.0885	66.7916	67.4947	68.1977	68.9008	69.6039
100. .	70.3070	. . .	. . .	. . .	. . .	. . .	. . .	. . .	. . .	. . .

Note 1: This table may be used for obtaining metric equivalents of quantities expressed in psi by multiplying the given values by 10^{-3}, that is, by moving the decimal point three places to the left.

Note 2: This table may be extended to values below 1 or above 100 ksi by manipulation of the decimal point and addition.

imum recommended clearances for oil-impregnated bearings used with ground steel shafting are listed in Table III.

Dimensional Tolerances for Plain Cylindrical Bearings. The data in Table IV are intended for bronze-base bearings with a 4 to 1 maximum length to inside diameter ratio and a 24 to 1 maximum length to wall thickness ratio, and for iron-base bearings with a 3 to 1 maximum length to inside diameter ratio and a 20 to 1 maximum length to wall thickness ratio. Bearings having greater ratios than these are not covered by the table.

Removing Oil. For small quantities the procedure outlined under Density (line 3) of the Materials Standard may be used. For large quantities an active oil solvent such as lead-free gasoline or trichlorethylene may be used. The bearings are immersed for several hours, and the solvent is changed 2 to 5 times, as required, during this period. After extraction the bearings are allowed to

Table VIII. Pressure and Stress Equivalents (Thousand pounds-

Conversion factors: 1 in. =
1 lbf =

ksi	0	1	2	3	4
					MN/m^2
0. . .	0.0000	6.8948	13.7895	20.6843	27.5790
10. . .	68.9476	75.8423	82.7371	89.6318	96.5266
20. . .	137.8951	144.7899	151.6847	158.5794	165.4742
30. . .	206.8427	213.7375	220.6322	227.5270	234.4217
40. . .	275.7903	282.6850	289.5798	296.4746	303.3693
50. . .	344.7379	351.6326	358.5274	365.4221	372.3169
60. . .	413.6854	420.5802	427.4749	434.3697	441.2645
70. . .	482.6330	489.5278	496.4225	503.3173	510.2120
80. . .	551.5806	558.4753	565.3701	572.2648	579.1596
90. . .	689.5281	627.4229	634.3177	641.2124	648.1072
100. . .	689.4757	. . .	. . .	. . .	. . .

Note 1: This table may be used for obtaining metric equivalents of quantities expressed in psi by multiplying the given values by 10^{-3}, that is, by moving the decimal point three places to the left.

Note 2: This table may be extended to values below 1 or above 100 ksi by manipulation of the decimal point and addition.

dry by evaporation. (Caution-observe common safety rules for fire and health hazards.)

Re-impregnation. Bearings may be re-impregnated by following the procedure outlined under Density (line 3).

Storage. Oil-impregnated bearings should be stored in non-absorbent containers to prevent oil loss. They should also be protected from dirt and dust – especially close-tolerance bearings.

Surface Finish. Where surface finish affects the function of a bearing, the purchaser may request a very smooth finish. However, due to the porous nature of powder metallurgy parts, normal stylus measurements with tracer type instruments will not measure the true finish of the surface. This is because the surface porosity is deeper than surface irregularities in the metal particles.

force per square inch to meganewtons per square meter)

0.0254 m (exactly)

4.448 221 615 260 5 MN (exactly)

5	6	7	8	9
34.4738	41.3685	48.2633	55.1581	62.0528
103.4214	110.3161	117.2109	124.1056	131.0004
172.3689	179.2637	186.1584	193.0532	199.9480
241.3165	248.2113	255.1060	262.0008	268.8955
310.2641	317.1588	324.0536	330.9483	337.8431
379.2116	386.1064	393.0012	399.8959	406.7907
448.1592	455.0540	461.9487	468.8435	475.7382
517.1068	524.0015	530.8963	537.7911	544.6858
586.0544	592.9491	599.8439	606.7386	613.6334
655.0019	661.8967	668.7914	675.6862	682.5810
...	...	...	...	...

The purchaser and manufacturer may agree on a surface finish specification and method of measurement.

Heat Treatment. Ferrous P/M parts containing 0.3% or higher combined carbon can be quench hardened for increased strength and wear resistance. The percentages of carbon and other alloying elements effectively combined in the material, and its density, determine the degree of hardening possible for any given quench condition. Surface hardness values of 500 to 650 Knoop, which are file hard, can be obtained readily by quench hardening.

Ferrous parts initially made without carbon can readily be carburized by standard means other than liquid salts. Low-density parts carburize throughout while high-density parts develop a distinct carburized case. Good process control is necessary to assure that specified carbon levels are not exceeded. Methods of case depth evaluation are covered by MPIF Standard 37. Very high density parts respond favorably to fused salt carbonitriding. In this method density must be high enough to prevent salt absorption into the pore structure.

Low and medium density P/M parts readily absorb salts, brines, and oils into the pore structure. Therefore, oil quench hardening is recommended in preference to other methods because pore-entrained quench oil is not corrosive.

Tempering for stress relief after quenching is further recommended for maximum strength and durability. During tempering, provision must be made to vent or disperse oil vapors generated by quench oil in the material pore structure.

Index

MIX
Papier aus verantwortungsvollen Quellen
Paper from responsible sources
FSC® C105338

If you have any concerns about our products,
you can contact us on
ProductSafety@springernature.com

In case Publisher is established outside the EU,
the EU authorized representative is:
Springer Nature Customer Service Center GmbH
Europaplatz 3, 69115 Heidelberg, Germany

Printed by Libri Plureos GmbH
in Hamburg, Germany